13.8

THE QUEST TO FIND THE TRUE
AGE OF THE UNIVERSE AND THE
THEORY OF EVERYTHING

JOHN GRIBBIN

ICON

First published in the UK in 2015 by Icon Books Ltd

This edition published in the UK in 2016 by
Icon Books Ltd, Omnibus Business Centre,
39–41 North Road, London N7 9DP
email: info@iconbooks.com
www.iconbooks.com

Sold in the UK, Europe and Asia
by Faber & Faber Ltd, Bloomsbury House,
74–77 Great Russell Street,
London WC1B 3DA or their agents

Distributed in the UK, Europe and Asia
by Grantham Book Services,
Trent Road, Grantham NG31 7XQ

Distributed in Australia and New Zealand
by Allen & Unwin Pty Ltd,
PO Box 8500, 83 Alexander Street,
Crows Nest, NSW 2065

Distributed in South Africa by
Jonathan Ball, Office B4, The District,
41 Sir Lowry Road, Woodstock 7925

Distributed in India by Penguin Books India,
7th Floor, Infinity Tower – C, DLF Cyber City,
Gurgaon 122002, Haryana

ISBN: 978-178578-108-7

Typeset in Dante by Marie Doherty

Printed and bound in the UK
by Clays Ltd, St Ives plc

Contents

About the Author

John Gribbin was born in 1946 in Maidstone, Kent. He studied physics at the University of Sussex and went on to complete an MSc in astronomy at the same university before moving to the Institute of Astronomy in Cambridge, to work for his PhD.

After working for the journals *Nature* and *New Scientist*, he has concentrated chiefly on writing books on everything from the Universe and the Multiverse to the history of science. His books have received science-writing awards in the UK and the US. His biographical subjects include Albert Einstein, Erwin Schrödinger, Stephen Hawking, Richard Feynman, Galileo, Buddy Holly and James Lovelock.

Since 1993, Gribbin has been a Visiting Fellow in Astronomy at the University of Sussex.

Acknowledgements

The University of Sussex provided me with a base to work from, and the astronomy group there provided many stimulating discussions on various aspects of astronomy. Beyond Sussex, Virginia Trimble of the University of California, Irvine kept us straight on history, and François Boucher of the Institut d'Astrophysique Paris kept us up to date on the Planck satellite discoveries. I am also grateful to the Alfred C. Munger Foundation for continued financial support.

List of Illustrations

34. Cosmic microwave background seen by Planck
The map shows tiny temperature fluctuations that correspond to regions of slightly different densities at very early times, representing the seeds of all future structure, including the galaxies of today.
ESA and the Planck Collaboration

35. Planck power spectrum of the cosmic microwave background
The dots are measurements made by Planck. The wiggly line represents the predictions of the 'Lambda–CDM' model of the Universe.
ESA and the Planck Collaboration

Introduction

The Most Important Fact

The Universe began. The origin of everything we see about us – stars, planets, galaxies, people – can be traced back to a definite moment in time, 13.8 billion years ago. The 'ultimate' question that baffled philosophers, theologians and scientists for millennia has been answered in our lifetime. It has taken almost exactly half a century, starting in the mid-1960s with the discovery of the cosmic microwave background radiation,[1] for the idea of a Universe of finite age to go from being a plausible hypothesis – but no more plausible than the idea of an eternal, infinite Universe – to being established as fact. The age of the Universe has been measured with exquisite precision using data from space observatories such as Planck. But accounts of this scientific triumph often overlook the fact that there is a second leg to the journey. The existence of this second leg is what makes the discovery of the beginning so compelling.

The most important thing we know in science is that our theory of the very small – quantum theory – agrees precisely with our theory of the very large – cosmology, aka the general theory of relativity. This is in spite of the fact that the two theories were developed entirely independently and that nobody has been able to unify these two great theories into one package, quantum gravity. But the fact that they separately give the 'right' answers to the same question tells us that there is something fundamentally correct about the whole of physics and, indeed, the whole scientific enterprise. It works.

What is that profound question? How do we know they agree? Because the age of the Universe calculated by cosmologists, *13.8 billion years*, is just a tiny bit older than the ages of the stars it contains, as calculated by astrophysicists. This is such a profound insight that it ought to be shouted from the rooftops; instead, it is taken for granted. I intend to redress the balance.

Recent events have highlighted the way in which the significance of this agreement has slipped under the radar. I was provoked into writing this book when, in the spring of 2013, data from the Planck satellite made headlines. The story trumpeted by the media was that 'the Universe is older than we thought'. This caused wry amusement amongst cosmologists. Although true, what the data told us is that the estimated age of the Universe had increased from 13.77 billion years to 13.82 billion years, an increase of less than half of one per cent (later revised down to 13.80 billion years). What is more astonishing about these data is that we know the age of the Universe to such a degree of accuracy. A generation ago (although even then we knew that there had been a beginning), we could only say that the Universe was somewhere between 10 and 20 billion years old. The precision of the new measurement is half of the most important fact – both in physics, which is the focus of this book, as well as in the wider world of thought. The philosophical and religious implications I leave for others to debate.

The ages of the oldest stars show that they are just a little bit younger than the Universe. If that doesn't sound impressive, imagine how scientists would feel if it were the other way round – if stars were measured as being older than the Universe! It would tell them that at least one of their two most cherished theories, quantum physics and the general theory of relativity, must be wrong.

In fact, we don't have to imagine how scientists would feel if stars were measured as being older than the Universe. The consensus I have just described has emerged since the end of the Second World War, which coincidentally means that it has emerged precisely during my lifetime and that I was not only a member of one of the teams that measured the age of the Universe but knew personally many of the people involved in this story. When I was a child, astronomers did indeed find that their estimates of the ages of stars came out bigger than their estimate of the age of the Universe. This was one of the underpinnings of the 'steady-state' model, which perceived the Universe as infinite in time and space, and essentially unchanging. I will explain how we got from the apparent conflict of the 1940s to the modern consensus, including the significance of the Planck results, and will make the importance of this consensus clear. But I will also set the scene by looking at the 'prehistory' of the subjects, cosmology and astrophysics, going back to the 19th century discoveries that pointed the way to an understanding of the nature of stars and the Universe – to the most important fact.

John Gribbin
1 June 2015

PART ZERO

Prologue

2.712

Taking the temperature of the Universe

Half a century ago, in 1965, American astronomers Arno Penzias and Robert Wilson announced that they had accidentally discovered a weak hiss of radio noise coming from everywhere in space. Although they were unaware of it at the time, this 'cosmic microwave background radiation' had been predicted, more than a decade earlier, by George Gamow and colleagues in the context of the Big Bang model of the Universe. Bizarrely, unknown to Penzias and Wilson, in 1965 another team of astronomers, headed by Jim Peebles, had also come up with the idea (and were also unaware of the Gamow team's work) and were building a detector to search for the radiation. When news of the discovery reached Peebles, he quickly interpreted it as evidence for the Big Bang, but even in their discovery paper Penzias and Wilson deliberately refrained from making this connection, because they favoured the rival steady-state model. Nevertheless, this publication marked the moment, dated almost to the day, when the idea of the Big Bang became the leading cosmological paradigm. The temperature of the background radiation today – 2.712 K, or −270.288°C – is an indicator of how hot the Universe was 'in the beginning' and is persuasive evidence that there was a beginning.

But Penzias and Wilson had no idea of the significance of their discovery at the time. They were working at the Bell Laboratories of the American Telephone and Telegraph Company (AT&T), using an antenna designed and built to test the feasibility of global communication via satellites. They were able to use the antenna, located at Crawford Hill in New Jersey, for purely scientific research because of the enlightened policy of AT&T allowing their Bell Labs scientists freedom to carry out such research alongside their practical investigations of ways to improve telecommunications.

Bell Telephone Laboratories came into existence as the research arm of AT&T on 1 January 1925. Just two years later, two Bell Labs researchers, Clinton Davisson and his assistant Lester Germer, confirmed the wave nature of the electron, a key development in quantum physics. As a result, in 1937 Davisson became the first Bell Labs scientist to receive the Nobel Prize. He would not be the last. The transistor was invented at Bell Labs, for which John Bardeen William Shockley and Walter Brattain shared the Nobel Prize in 1956. By the early 1960s, the Bell Laboratories were widely recognised as centres of scientific excellence, where many young researchers were eager to work.

One of those young researchers was Arno Penzias. He had been born into a Jewish family, the son of a Polish (but German-born) father and a German mother, in Munich, on 26 April 1933, the same day that the Gestapo was formed. As the eldest child in a comfortable middle-class family, the troubles in Germany in the 1930s passed him by until 1938, by which time the Nazis were rounding up Jews who did not hold German passports and sending them into Poland. The Polish authorities had almost as great an antipathy towards the Jews as the Nazis had, and effectively closed the border to the exodus on 1 November 1938. The train on which the Penzias family were passengers arrived a couple of hours later, and they were sent back to Munich, where Arno's father was given six months to get the family out of Germany or face the consequences. At the age of six, Arno was put in charge of his younger brother and sent on a train to England. The boys' parents managed to get separate visas a little later and escaped just before war broke out. With great foresight, months before, Mr Penzias had bought tickets for New York, and the family travelled there by ocean liner in December 1939, spending Christmas and New Year on board.

Although life as refugees in America was financially much harder than it had been in Germany, as Penzias put it in his Nobel Prize autobiographical note: 'it was taken for granted that I would go to college, studying science'. The only affordable option was City College of New York, where Arno met his future wife, Anne. When they had arrived in New York, the children had taken American first names, Arno becoming 'Allen' and his brother Gunter becoming 'Jim'. But Anne already knew an Al, and called Penzias 'Arno' to avoid confusion. He got his name back, and took to signing himself 'Arno A. Penzias'.

Arno and Anne married in 1954, the year he graduated from City College, and after two years in the Army Signal Corps he moved to Columbia University, completing a PhD in 1961 under the supervision of Charles Townes, who would receive the Nobel Prize for his work on masers and lasers in 1964. Townes had worked at Bell Labs from 1939 to 1947. It was Townes who introduced Penzias to Bell Labs, where he was offered a job in 1961. In the long term, Penzias intended to use the horn antenna at Crawford Hill for radio astronomy work, but at the time it was still reserved for use with satellites, notably Telstar (designed by Bell and due for launch in 1962), so he worked on another project. It turned out that the horn antenna was not needed for the Telstar work after all, and it became available for radio astronomy just about the time a second radio astronomer, Robert Wilson, joined Penzias at Bell. They began working together early in 1963.

Wilson was slightly younger than Penzias, having been born in Houston, Texas on 10 January 1936. His father worked in the oil industry, on the exploration side, but had a hobby repairing radios, which gave Robert a basic grounding in electronics. He passed through the school system as a good but not outstanding student, then moved on to Rice University in 1953, 'having barely been admitted', according to his Nobel autobiography. He enjoyed the courses and 'the elation of success' so much that he graduated with honours, moving on to the California Institute of Technology (Caltech) in 1957 for a PhD in physics with no clear idea of what kind of research to do. There, he took a course on cosmology given by Fred Hoyle, which made him an enthusiast for the idea of a steady-state universe (more of this later), but more significantly he followed up a suggestion made by David Dewhirst (like Hoyle, a visitor from Cambridge) to work in radio astronomy. Before doing so, he went back to Houston for the summer of 1958, where he married Elizabeth Sawin.

For his research project, Wilson made a radio map of the Milky Way, using a new telescope at the Owens Valley Radio Observatory; the work involved the ideal mix, for him, of electronics and physics. His thesis was submitted in 1962. Wilson had originally been supervised by John Bolton, an Australian who played a large part in the construction of the telescope, and when Bolton returned to Australia the supervisory role was taken over by Maarten Schmidt. Wilson 'developed a good feeling toward Bell Labs' during this work, when they developed a pair of maser amplifiers for use at the Owens Valley telescope, and he had also heard about the new horn antenna. He joined the Crawford Hill team in 1963, where it clearly made sense for him to work on a joint project with Arno Penzias, the only other radio astronomer there, rather than working separately. The collaboration was to endure – when financial cuts reduced the funding available for radio astronomy at Crawford Hill to one full-time researcher, they agreed to work for half their time on radio astronomy and to devote the other half to more immediately practical work. But that happened after the discovery for which they won the Nobel Prize.

The shape of the horn antenna is designed to minimise interference from the ground and to provide the best possible measurement of the strength of radio noise (like light, part of the electromagnetic spectrum) coming from different places in space, primarily artificial satellites but also natural objects such as stars and clouds of gas. The strength of this radio noise is measured in terms of temperature, calibrated by the temperature of radiation emitted by a so-called 'black body'. This counter-intuitive term for a radiating object came about because objects that are perfect absorbers of electromagnetic radiation (hence black) are also, when heated, perfect emitters of radiation (see Chapter One). The nature of this radiation depends precisely on the temperature of the radiating object.

Scientists measure temperature in degrees Kelvin, denoted by K (without a degree sign, °). Each degree is the same size as one degree Celsius, but 0 K is the absolute zero of temperature, the lowest possible temperature, which corresponds to $-273.15\,°$C. In round numbers, the average temperature of the surface of the Earth is about 300 K. But the superb design of the horn antenna meant that the interference from the ground picked up by the radio telescope was less than 0.05 K. In order to do justice to the antenna, before they began astronomical observations Penzias and Wilson wanted to build a receiver, the electronic business end of the telescope (a radiometer), which was equally sensitive, or at least as sensitive as they could possibly make it.

The amplifiers used in the receiver (similar to the ones Wilson had used in California) were cooled to 4.2 K using liquid helium, and Penzias devised a 'cold load', itself cooled by liquid helium to about 5 K, to calibrate the system. By switching the antenna from observations of the cold load to observations of the sky, they could measure the apparent temperature of the Universe (expected to be zero K) then subtract out known factors, such as the interference from the atmosphere above and the radiometer. What was left, they thought, would be noise due to the antenna itself, which they could then eliminate by whatever means proved appropriate (polishing it, maybe). Of course, what they hoped was that there would be no residual noise, that the telescope was working fine, and that they could get on with some radio astronomy.

In fact, something similar to this calibration had already been done, using slightly less accurate technology, and without the all-important cold load, by the engineers who built the horn antenna, to check that it was sensitive enough to do the job it had been designed for. One of them, Ed Ohm, had published their results in the *Bell System Technical Journal* in 1961. He reported that the temperature measured by the telescope when pointed at the sky was 22.2 K, with an uncertainty of plus or minus 2.2 K, meaning that it could be anything in the range from 20 to 24.4 K. His team's calculation of the amount of noise in the system from the atmosphere, the residual heat of the radiometer and so on came out as 18.9 K, plus or minus 3 K, making anything from 15.9 to 21.9 K possible. Taking the middle of each range at face value, and subtracting one from the other, you would be left with 3.2 K as the temperature of the sky. But within these ranges of uncertainties, the two sets of numbers agreed with one another. So Ohm concluded that 'the most likely minimum system temperature' was therefore 21 ± 1 K. But as Penzias and Wilson refined their system, the errors became smaller and a gap grew between the expected measurements and the actual measurements. It soon became clear that the radiation coming from the antenna into the receiver was at least 2 K hotter than they could explain.

The pair did everything they could think of to remove any sources of interference in the antenna, including cleaning out the layer of droppings that had accumulated in it from a pair of nesting pigeons and sticking shiny aluminium tape over all the riveted joints; nothing made much difference. The mystery of the 'excess antenna temperature' continued to baffle them throughout 1964, putting their whole radio astronomy research project in jeopardy. But they still had time for other things, and in December 1964 Penzias made the acquaintance of a fellow radio astronomer, Bernard Burke, of the Massachusetts Institute of Technology (MIT), at a meeting of the American Association for the Advancement of Science, held in Washington, DC. Three months later, during a telephone conversation about something else, Penzias mentioned to Burke the continuing problems with the antenna noise. Burke told Penzias that he had heard that a team headed by Jim Peebles and Robert Dicke, working at Princeton University (just a half-hour drive away from Crawford Hill), was working on a project of their own which just might have some bearing on the problem. After talking things over with Wilson, Penzias called Dicke, who happened to be in a meeting with his colleagues – Peebles and two junior researchers, Peter Roll and David Wilkinson. Dicke listened intently to what Penzias had to say, occasionally making comments. As he put the phone down, he turned to his colleagues and said: 'Well, boys, we've been scooped.'[2]

Unknown to Penzias and Wilson, the Princeton team was investigating the idea that the Universe had expanded from a hot, dense state which had left it filled with cold background radiation, radio noise in the microwave band. They were actually building a small radio telescope to look for this radiation. The next day, they made the 30-mile trip to meet Penzias and Wilson and to check out the radio telescope. They were quickly convinced that the Bell researchers had found this 'relict' radiation, that the 'excess' temperature was nothing to do with the antenna itself but was actually the temperature of the Universe at large. Penzias and Wilson were not so sure, not least because they favoured the steady-state idea, which said that on the largest scales the Universe is essentially eternal and unchanging. But they were relieved to be offered any kind of a scientific explanation for their measurements.

What, though, was this explanation? Dicke's idea might be most succinctly described as 'the Big Bang, but not as we know it'. Dicke, born in 1916, was from an older generation than Penzias, Wilson and his junior colleagues at Princeton. He had worked on radar during the Second World War and had developed an instrument known as a Dicke Radiometer to study exactly the kind of microwave radiation that would later attract the attention of Penzias and Wilson. Indeed, in 1946, while using such an instrument to study the radiation from the Earth's atmosphere, he found that any 'noise' coming from straight overhead (that is, from space) must correspond to a temperature below 20 K; but at the time he was not thinking about cosmology, and by 1965 he had completely forgotten that he had made this measurement. He had become interested in the background radiation again because of a puzzle involving the origin of the elements – a recurring theme of the various strands of research described in this book.

By the middle of the 1940s, it was clear, as I shall explain in Chapter One, that most of the visible matter in the Universe is in the form of hydrogen and helium. Roughly 75 per cent of the stuff in bright stars and galaxies is hydrogen, some 24 per cent helium, and about 1 per cent everything else, including the stuff that planet Earth and your body is made of. Hydrogen is the simplest element, and each atom of hydrogen consists of a single proton accompanied by a single electron. Assuming that this is the basic building block of matter, astrophysicists puzzled over where everything else had come from.

The first person to apply cosmological ideas to try to calculate how the other elements had formed was George Gamow, a Russian émigré physicist then working at George Washington University in Washington, DC. Gamow was one of the first physicists to espouse the idea fully – based on the then-new evidence that the Universe is expanding (see Chapter Six) – that the Universe had been born in a hot, dense state, now known as the Big Bang. Gamow guessed that the Universe might have started out as a hot, dense gas of neutrons. These neutral particles are unstable and quickly decay, each one breaking down into a single proton and a single electron, giving a supply of hydrogen. If the conditions in the Big Bang were hot enough and dense enough, protons (hydrogen nuclei) could be forced together to form nuclei of deuterium (heavy hydrogen), a process known as fusion, with further collisions building up nuclei of helium, each of which consists of two protons and two neutrons. Gamow gave a graduate student, Ralph Alpher, the task of carrying out the calculations to see just how effective this process would be, and together they found that, although it would indeed be easy to make helium in this way, it would be very difficult to build up heavier elements before the expanding Universe cooled and fusion reactions stopped. Unabashed, Gamow, a larger-than-life character who never doubted his own ability, pointed out that the theory explained where 99 per cent of the visible Universe came from, and that the rest was a detail which could be left for other people to work out.

The work formed the basis of Alpher's PhD thesis and was adapted into a scientific paper published in the journal *Physical Review* in 1948. Gamow, an inveterate joker, decided to add the name of his friend Hans Bethe as co-author of the paper (without Bethe's knowledge) so that it would be signed Alpher, Bethe, Gamow, in tribute to the Greek alphabet (alpha, beta, gamma). Alpher was not exactly delighted at this dilution of credit for his first significant piece of work but had little choice in the matter. To this day it is known as the 'alpha, beta, gamma' paper. But at least his name came first. It was a key step in cosmology simply because it introduced the idea that you could do real calculations involving the Big Bang. But it left unanswered the question of the origin of all the elements except hydrogen and helium.

The puzzle of the origin of the elements (nucleosynthesis) was one of the reasons why an alternative to the Big Bang idea, known as the steady-state model, was put forward by Hermann Bondi, Tommy Gold and Fred Hoyle, also in 1948. The basic idea was that, although the Universe is expanding – so that the islands of stars known as galaxies are getting farther apart from one another – it has not expanded out of a hot, dense state a definite time ago but has always looked much the same. As it expands new matter, in the form of hydrogen atoms, appears in the gaps between galaxies and is incorporated into new stars and galaxies. Nucleosynthesis then takes place inside stars. This is a much slower process than the kind of Big Bang nucleosynthesis envisaged by Gamow and his colleagues, but since the steady-state model says that the Universe is infinitely old, time is not a problem. As we shall see, Hoyle, in particular, was instrumental in developing an understanding of stellar nucleosynthesis, and for a time in the late 1950s he was able to dismiss the Big Bang idea as unnecessary (incidentally, he coined the name 'Big Bang' in a BBC radio broadcast). But it turned out, as Hoyle himself found, that although stellar nucleosynthesis can indeed explain the 1 per cent, it cannot explain the origin of all the helium in the Universe. You need both Big Bang nucleosynthesis and stellar nucleosynthesis to account for all the stuff of the visible Universe. But that is getting ahead of our story.

Dicke didn't like the idea that all the stuff of the Universe could have been made in a fraction of a second in a Big Bang; nor did he like the idea that matter is being continuously created in the spaces between galaxies. There was, though, a third alternative, known as the cyclic universe. In this picture, the amount of matter in the universe stays the same, but after a phase of expansion there is a phase of collapse, leading in to a hot, dense state just like the

Big Bang, from which the universe bounces out, Phoenix-like, for another cycle.*

By the 1950s, it was clear that there are two families of stars in a galaxy like our own Milky Way, so-called Population I and Population II. Population II are old stars that contain relatively little in the way of heavy elements (to astronomers, all elements heavier than helium are known as 'metals'). They are almost entirely hydrogen and helium. Population I are young stars that contain relatively high proportions of heavy elements (metals). The inference is that the younger stars have been made from material recycled from previous generations of stars and enriched (or dirtied) with metals – clear evidence for stellar nucleosynthesis. But for the cyclic (or oscillating) universe model to work, Dicke realised, the dense phase would have to be hot enough to clean up the universe by breaking all of these metals back down into hydrogen and helium. This led him to the idea that the Universe we see around us has indeed expanded from a hot, dense state, even if it was not a unique Big Bang. Sometime in 1964 he suggested to Jim Peebles, a colleague who had recently completed his doctorate, that Peebles might calculate the temperature required to do the job and the temperature of the resulting leftover radiation today. Peebles' rough calculation suggested that the Universe today should be filled with a sea of microwave radiation at a temperature below 10 K, and Roll and Wilkinson were setting out to find this radiation when the call came from Penzias.

The upshot of the meeting between the two teams was that they produced a pair of papers which were published alongside each other in the July 1965 issue of the *Astrophysical Journal*. The paper by Dicke, Peebles, Roll and Wilkinson came first, setting out the theory of leftover radiation from a hot early Universe; then came

---

* Dicke's oscillating universe model was actually a little more complicated than this, but since it was wrong I won't go into the details.

the paper by Penzias and Wilson, prosaically titled 'A Measurement of Excess Antenna Temperature at 4,080 mc/s'. It made no mention of the possible significance of the discovery except for the sentence: 'A possible explanation for the observed excess noise temperature is the one given by Dicke, Peebles, Roll and Wilkinson in a companion letter in this issue.' They were not yet ready to abandon the steady-state model! 'We thought,' said Wilson in his Nobel Lecture, 'that our measurement was independent of the theory and might outlive it.' In fact, according to Dicke: 'Penzias and Wilson weren't even going to write a paper at all until we told them we were writing one.'[3] But in 1978, after many measurements at a wide range of wavelengths, by many teams of astronomers, had confirmed that what they had discovered really was the leftover radiation from the Big Bang itself, now with a temperature of 2.712 K, Penzias and Wilson shared the Nobel Prize for the discovery. It has been suggested that Penzias and Wilson would have been happy simply to have news of their discovery added to the Dicke, Peebles, Roll and Wilkinson paper, with their names as fifth and sixth authors.* If so, the Nobel Prize would probably have gone to Dicke. But don't feel too sorry for him; there are other candidates worthy of your sympathy in this story.

Ralph Alpher didn't stop thinking about the Big Bang after he completed his PhD. By then, he was working with another of Gamow's protégés, Robert Herman, following up another of Gamow's insights. Gamow had a happy, but sometimes (to his colleagues) infuriating, knack of coming up with profound insights based on incomplete, or just plain wrong, reasoning. In 1948, he came up with an idea that Penzias described as 'wrong in almost every detail', but which contained a profound truth.[4] He realised

---

* The suggestion came either from Penzias or from Wilson, but I have mislaid the source.

that although the Big Bang had to be hot in order for nuclear fusion to work, it could not be too hot, or energetic photons (particles of light) would break helium nuclei apart as fast as they had formed. This sets a limit of roughly a billion degrees (10^9 K) at the end of the fireball phase that produced the helium, regardless of what conditions were like at even earlier times. Alpher and Herman took this idea and refined it, making it correct in almost every detail, and extended it by calculating that the leftover radiation from this fireball should still fill the Universe today, with a temperature of a few K; this result was published in 1948 in a short note in one of the most widely read scientific journals, *Nature*.[5] They concluded that: 'the temperature in the universe at the present time is found to be about 5 K'.

The suggestion is often attributed to Gamow himself, but this is wrong. According to Alpher and Herman: 'although our good friend and colleague Gamow did not at first believe that our prediction of 5 Kelvin was meaningful, useful, or amenable to observation, and several years elapsed before he took it seriously, thereafter he wrote about the subject in a number of papers.'[6] Gamow was also a great populariser of science and wrote about the idea in his books, leading to the widespread notion that he thought of it – an example, as Alpher and Herman point out, of the Matthew Effect.* In *The Creation of the Universe* (1952), for example, Gamow wrote: 'we find $T_{present} = 50$ degrees absolute', an overestimate based on a typically Gamowian arithmetical slip, but still something that ought to have made scientifically aware readers sit up and take notice. It is astonishing that Dicke and his colleagues were unaware of the work of Alpher and Herman prior to 1964, not least because Dicke had been working with microwave equipment in the 1940s. If he had read

---

* 'For unto every one that hath shall be given, and he shall have abundance: but from him that hath not shall be taken even that which he hath.' (Matthew 25:29, King James Version)

the paper by Alpher and Herman, even with the technology of the time (and a suitable cold load) he would have been able to detect the microwave background, and Alpher and Herman would have received due credit. Even stranger, both Wilson and Wilkinson have said that their early interest in science was stimulated by reading Gamow's books, but the story of the background radiation seems to have passed them by.[7]

Naturally, Gamow, Alpher and Herman were deeply upset when the discovery made headline news without any mention of them – the first they knew of it was a front-page story in *The New York Times*. The resulting recriminations have been well documented by John Mather and John Boslough, two later players in the cosmic background game; there is no need to elaborate on them here.[8] But a couple of other missed opportunities are worth mentioning.

As I have explained in my book *In Search of the Big Bang*, the string of missed opportunities to identify the background radiation goes right back to the early 1940s and studies of the spectra of starlight that has passed through clouds of interstellar material, a mixture of gas and dust. The way this light is absorbed, leaving lines in the spectrum, can reveal the temperature of those clouds, and through studies of a particular feature associated with molecules of cyanogen, Andrew McKellar of the Dominion Astrophysical Observatory in Canada inferred that these clouds had a temperature of between 2 and 3 K. The result was well known to astronomers, but nobody realised that the clouds were being kept at this temperature by being bathed in the background radiation, as if they were in a very cool microwave oven.

My favourite story of how people who should have known better missed the implications involves Fred Hoyle and George Gamow. In 1956, Hoyle was visiting La Jolla, California, where Gamow was also on a short-term visit, driving around in a brand-new white Cadillac convertible (an archetypal Gamow vehicle). At that time, Gamow, the prime mover of the Big Bang idea, was promoting the idea that the Universe was filled with a sea of radiation with a temperature of about 5 K, and Hoyle, the prime mover of the steady-state model, thought that there should be no such radiation. They had plenty to talk about. Hoyle told the story in an article in *New Scientist* in 1981:

> There were times when George and I would go off for a discussion by ourselves. I recall George driving me around in the white Cadillac, explaining his conviction that the Universe must have a microwave background, and I recall my telling George that it was impossible for the Universe to have a microwave background with a temperature as high as he was claiming, because observations of the CH and CN radicals by Andrew McKellar had set an upper limit of 3 K for any such background. Whether it was the too great comfort of the Cadillac, or because George wanted a temperature higher than 3 K whereas I wanted a temperature of zero K, we missed the chance [...] For my sins, I missed it again in exactly the same way in a discussion with Bob Dicke at the 20th Varenna summer school on relativity in 1961. In respect of the microwave background, I was evidently not 'discovery prone'.[9]

Nor was anyone else, except Penzias and Wilson! And it seems Gamow really only had himself to blame for being scooped by the Bell team.

By 1964, even Hoyle had begun to have doubts about the steady-state model, at least in its simplest form. Because it had proved impossible to make the required amount of helium inside stars, he had been investigating the possibility of making the helium somewhere else, if not in a single Big Bang then in a series of 'little bangs' spread out across the Universe. He developed the idea with a junior colleague, Roger Tayler, and together they calculated that such a series of events would also have produced a sea of background radiation – Hoyle, of course, knew all about the work of Alpher and Herman, but had arrived at the same conclusion by a different route. Somehow, though, even in 1964 he did not link this with McKellar's observations. In the first draft of the paper that Hoyle and Tayler prepared for publication, they included a prediction of the cosmic background radiation; Hoyle took this out before publication, although Tayler, as he told me much later, wanted it kept in.

But the closest near miss in the saga of the discovery (or non-discovery) of the background radiation came from Russia. In a flurry of work mostly being carried out over a few months and published in 1964, Soviet researchers put together all the pieces of the puzzle except one. Yakov Borisovich Zel'dovich, one of the towering figures of Soviet-era science, had carried out calculations similar to those of the Gamow team and concluded, like them, that the Universe must have started in a hot Big Bang which had left a relict radiation with a temperature of a few K. He even knew of Ohm's paper in the *Bell System Technical Journal*, but misunderstood Ohm's conclusion, as we shall see. A less exalted Soviet astronomer, Yuri Smirnov, calculated a temperature for the background radiation in the range of 1 K to 30 K, and, jumping off from his calculations, Andrei Doroshkevich and Igor Novikov wrote a paper in which they pointed out that the antenna best suited for the job of detecting this radiation would be the horn antenna on Crawford Hill. The reason why none of the Russians realised that Ohm had already found this radiation was because something had been lost in translation. Ohm's paper said that he had measured the temperature of the sky to be about 3 K. He meant that after he had subtracted out all other possible sources of radio noise, he was left with a 3 K background. But by coincidence, the effect of the atmosphere on the antenna, one of the things that Ohm had subtracted out, is also about 3 K. The Russians thought that Ohm meant that he had measured this temperature, so they subtracted it out again and were left with nothing. Today, such confusion would quickly be sorted out by email, but communications between scientists in the Soviet Union and the United States were severely restricted in the early 1960s.

In spite of all the false starts and misunderstandings, though, the cosmic microwave background radiation *was* discovered. Over the following years and decades, it was studied in increasing detail, and some of the fruits of those investigations will be described in Part Two of this book. The essential point is that this radiation, with a temperature of 2.712 K, tells us that the Universe as we know it had a definite beginning a finite time ago. But when? This is where the story really begins.

PART ONE

How Do We Know the Ages of Stars?

1 2.898

Prehistory: Spectra and the nature of stars

The positivist philosopher Auguste Comte wrote in 1835 that: 'there is no conceivable means by which we shall one day determine the chemical composition of the stars'. Unknown to him, the first step towards that understanding had, in fact, already been taken, and the process would be completed not long after his death in 1857.

Locating lines

That first step had been taken as early as 1802, when Comte was only four years old, by the English physician and scientist William Hyde Wollaston. In spite of being partially blind from 1800 onwards, Wollaston, one of the leading scientists of his day, made many contributions to optics. The 1802 discovery came when, following in the footsteps of Isaac Newton, he was studying the spectrum of sunlight passed through a slit to make a narrow beam and then through a glass prism to spread the beam into a rainbow pattern, the solar spectrum. He noticed that the colours were broken up by dark bands, and Wollaston counted two in the red part of the spectrum, three in the green and two in the blue-violet region. Wollaston mistakenly thought that these were simply gaps between the colours and did not pursue his investigation of the phenomenon. But his discovery triggered the interest of other researchers, most notably the German Joseph von Fraunhofer, who was able to produce much more detailed spectra in the second decade of the 19th century and eventually identified 574 separate lines. Today, all the dark lines in the solar spectrum, even more than he counted, are known as Fraunhofer lines. A short section of the spectrum has lines packed together, giving an appearance superficially rather like the lines of a barcode. But what causes them?

A partial answer came from the work of Robert Bunsen and Gustav Kirchhoff in Germany, in the 1850s and 1860s. Bunsen's name is known to everyone who has studied chemistry thanks to the eponymous burner – although in fact this kind of burner was invented by Michael Faraday and the design improved by Bunsen's assistant, Peter Desaga, who used the name of the more famous Bunsen, rather than his own, in marketing the improved device. But what matters here is not who invented the Bunsen burner, but what Bunsen and Kirchhoff did with it.

Early in the 1850s, the city of Heidelberg had been piped for the distribution of inflammable gas derived from coal to households and businesses – and to the scientific laboratories of the university. This was the inspiration for Bunsen's work with the burner that now bears his name. The burner combines oxygen and the inflammable gas in a controlled way that produces a clear flame ideal for use in the 'flame test', by which substances are identified by the colour they give to a flame. Bunsen originally used coloured filters to calibrate these observations, but Kirchhoff pointed out that it would be possible to make a more detailed analysis using spectroscopy. Together, they built an apparatus which included a narrow slit for the light to pass through, a collimator to narrow the beam, a prism to spread the light out into a rainbow pattern and an eyepiece, like that of a microscope, to view the spectrum. Although Fraunhofer had used a prism and eyepiece combination in his work, this was the first time all these components had been assembled together in a single instrument – a spectroscope.

The Heidelberg team knew that when different substances were put in the clear flame of such a burner they burned with different colours. A trace of sodium, for example, makes the flame yellow, while copper colours the flame green/blue. So they analysed the light from these flames using spectroscopy. They found that each element, when hot, produced bright lines in the spectrum at precise wavelengths – in the yellow part of the spectrum for sodium, in the green/blue part of the spectrum for copper, and so on. (The yellow sodium lines were also known to Fraunhofer, who had used them to test the optical properties of glass, which had led to his investigation of the solar spectrum.) The German team soon realised that any hot object produces distinctive lines in the spectrum. One evening, from their laboratory in Heidelberg they were able to analyse the light from a major fire in Mannheim, some ten miles away, and identify lines produced by the presence of strontium and barium in the blaze.

A few days later, Bunsen and Kirchhoff were walking along the Neckar River, which flows through Heidelberg, and discussing what they had seen in the fire. According to legend, Bunsen remarked to Kirchhoff something along the lines of: 'If we can determine the nature of substances burning in Mannheim, we should be able to do the same thing for the Sun. But people would say we have gone mad to dream of such a thing.'

Nevertheless, they turned their attention to the spectrum of the Sun and found that many of the dark lines identified by Fraunhofer were in the same part of the spectrum – at precisely the same wavelengths – as the bright lines produced by various elements when heated in the lab. The natural implication was that these elements are present in the outer layer of the Sun, but they are cooler than the layer below so as the light from the hot interior passes through this region they *remove* light from the spectrum at specific wavelengths instead of adding bright lines to it. Kirchhoff in particular developed this understanding of what was going on. Nobody at that time knew how the lines were produced – that would have to wait for the development of the quantum theory of atomic structure in the 20th century. But even without that understanding, in the 1860s it was now possible to find out what the Sun was made of – and, applying the same technique, what the stars were made of. Referring to their conversation by the river, Kirchhoff is said to have told his colleague: 'Bunsen, I have gone mad.' To which Bunsen replied: 'So have I, Kirchhoff!'[10] Kirchhoff's discovery was presented to the Prussian Academy of Sciences in Berlin on 27 October 1859, now regarded as the date on which the discipline of astrophysics was born (although it was not given that name until 1890).

It had taken a mere three decades to prove Comte wrong. Well, not quite. In the remaining decades of the 19th century, astronomers were able to identify the presence of many elements also found on Earth in the spectrum of the Sun and, with less detail, in the stars. The natural assumption they made was that the overall composition of the Sun was rather like the overall composition of the Earth. But this turned out to be wrong. Stars are much simpler than that, and we now know that they (the Sun included) are mostly composed of hydrogen and helium, with just traces of the other elements. But at the beginning of the 1860s, nobody even knew that there *was* such a thing as helium. Its discovery marked the coming of age of solar – and stellar – spectroscopy.

Hunting helium

The leading light in the discovery of helium was the British astronomer Joseph Lockyer, who was, in the 1860s, an enthusiastic amateur observer of the Sun. (His job was as a clerk at the War Office in London.) He quickly picked up on the developments in spectroscopy being made by Bunsen and Kirchhoff, and applied those techniques to his solar studies. Using spectroscopy, he showed that the darkness of sunspots is caused by the presence of relatively cool gas near the surface of the Sun absorbing light from the hotter gas below it. His greatest achievement came on 20 October 1868, when he was able to analyse light from the outer layers of the Sun with a new spectroscopic instrument.

These observations followed hot on the heels of a spectroscopic study of the outer layers of the Sun during an eclipse visible from India on 18 August that year. The observations – the first eclipse studied since the suggestion by Kirchhoff that the Fraunhofer lines correspond to the presence of different chemical elements in the Sun – were made by the French astronomer Pierre Janssen. At that time, with the Moon blocking out the bright light from the surface of the Sun itself, he could detect lines in the spectrum of the material just above the surface. He noticed bright lines in the spectrum of this layer of the atmosphere of the Sun, known as the chromosphere, including a bright yellow line with a wavelength later measured as 587.49 nanometres, close to the lines associated with sodium. The spectral lines were so bright that Janssen realised that they could be observed even without an eclipse, and he made more observations before returning to Europe.

On 20 October of the same year, unaware of Janssen's work, Lockyer used his new spectroscope to observe the solar atmosphere and found same yellow line. With impressive speed, both Janssen's and Lockyer's discoveries were presented to the French Academy of Sciences on 26 October 1868. But it was Lockyer who soon took things a step farther by claiming that the line must be associated with a previously unknown element, which he called helium, from the Greek word for the Sun, *Helios*.

This was a controversial claim. Many scientists preferred the idea that the line was associated with hydrogen subjected to extreme conditions of temperature and pressure. But in 1895 the physicist William Ramsay found that a previously unknown gas released by uranium produced a bright yellow line near to the sodium lines in the spectrum. He initially called this gas krypton, but when his colleague William Crookes pointed out that the line was in exactly the same place as the one found in the solar spectrum by Lockyer and Janssen he realised that it was, in fact, helium. (He later used the name krypton for another gas; nothing to do with Superman.) In effect, spectroscopy had predicted the discovery of helium on Earth, 27 years in advance.

By then, Lockyer had become a professional astronomer. In 1869 he was one of the founders of the scientific journal *Nature*, which he edited for the first 50 years of its existence, and in 1890 he was appointed Director of the Solar Physics Observatory in South Kensington, where he stayed until he retired in 1911. He was knighted in 1897, not least because of his discovery of helium.

As the discovery of helium shows, progress in astronomy proceeded on a broad front following the development of stellar spectroscopy, aided by new technological advances, not least photography, which, among other things, made it possible to keep a permanent record of stellar spectra that could be studied at leisure and compared with other spectra. But it makes sense here to jump forward to the 1920s and the next step towards an understanding of the composition of the stars, before doubling back to look at some of the other developments concerning their age.

Hunting hydrogen

That step was taken, appropriately, by someone born in 1900, with the new century. She also happened to be a woman, and it was still unusual for a woman to become a leading scientist in those days.

Cecilia Payne won a scholarship to Newnham College, Cambridge (the only way she could have afforded a university education) in 1919. She studied botany, physics and chemistry, but also attended a talk by Arthur Eddington about the eclipse expedition on which he had famously 'proved Einstein right' by measuring the way light from distant stars is bent by the Sun. This fired her interest in astronomy, and she visited the university's observatory on an open night, asking the staff so many questions that Eddington took an interest and offered her the run of the observatory library, where she read about the latest developments in the astronomical journals.

After completing her studies (as a woman, she was allowed to complete a degree course but could not be awarded a degree; Cambridge did not award degrees to women until 1948) Payne looked for a way to pursue this interest. There was no chance of a career in research in England, where the only job opportunities for women scientists were in teaching, but through Eddington she met Harlow Shapley, from Harvard, on a visit to England. He offered her the chance to work for a PhD on a graduate fellowship (even though, technically, she was not a graduate), and in 1923 she left for the United States. Just two years later, she produced a brilliant thesis and became the first person to be awarded a PhD by Radcliffe College (also the first for work carried out at Harvard College Observatory). In it, she established that the Sun is mainly made of hydrogen. But, in a sign of the times, the idea was not fully accepted until two male astronomers independently came to the same conclusion.

Payne's study of the solar spectrum made use of the then-recent discovery by the Indian physicist Meghnad Saha that part of the complication of the pattern of lines in a stellar spectrum (or the Sun's Fraunhofer lines) was a result of different physical conditions in different parts of the atmosphere of a star. By the 1920s, physicists knew (as, of course, Bunsen and Kirchhoff had not) that atoms are composed of a tiny central nucleus, with one or more electrons at a distance from the nucleus. Dark lines in a spectrum are produced when an electron absorbs a specific wavelength of light, moving to a higher energy level within the atom, and bright lines are produced when an electron drops down from one energy level to another and emits radiation (in the form, we would now say, of a photon). An atom that has lost one or more of its electrons is called an ion, and the spectra of ions are correspondingly different (in a way that can be calculated) from those of the 'parent' atoms. Payne measured the absorption lines in stellar spectra and showed how the temperature (in particular) and pressure in the atmosphere of a star affects the ionisation of the atoms there. This makes for a more complicated pattern of lines than if all the atoms were in their un-ionised state.* The spectra of stars differ from one another not because they are made of different things, but because of different amounts of ionisation in their atmospheres.

Payne's great achievement was to unravel this complicated pattern of hundreds of Fraunhofer lines and work out what proportion of different elements in different stages of ionisation had to be present to account for the observations. Some idea of the difficulty of her task can be gleaned from the fact that her thesis was later described by the astronomer Otto Struve as 'the most brilliant PhD thesis ever written in astronomy'. She worked out the proportions of eighteen elements in the Sun and stars, discovering that they all had nearly the same composition. But the big surprise was that according to her analysis the Sun and stars are made almost entirely of hydrogen and helium. If she was correct, everything else put together made up only 2 per cent of the composition of our nearest star, and of all stars. Most of the matter in the Universe was in the form of the two lightest elements, hydrogen and helium. This was almost literally unbelievable in 1925. Payne believed her results were correct, but when Shapley sent a draft of her thesis to Henry Norris Russell at Princeton for a second opinion, he replied that the result was 'clearly impossible'. On Shapley's advice, she added a sentence to the thesis saying that: 'the enormous abundance derived for these

---

* I am always careful to use the hyphen in the word 'un-ionised' since Isaac Asimov once pointed out to me that the way to distinguish between a scientist and a politician is to ask them how to pronounce the word 'unionised'.

elements [hydrogen and helium] in the stellar atmospheres is almost certainly not real'. But with the thesis accepted and her doctorate awarded, she wrote a book, *Stellar Atmospheres*, which began to persuade astronomers that the results were almost certainly real.

The change of mind was aided by the independent confirmation of Payne's results by other astrophysicists. In 1928, the German astronomer Albrecht Unsöld carried out a detailed spectroscopic analysis of the light from the Sun; he found that the strength of the hydrogen lines implied that there are roughly a million hydrogen atoms in the Sun for every atom of anything else. A year later, the Irish astronomer William McCrea confirmed these results using a different spectroscopic technique.* What this shows, more than anything, is that although Cecilia Payne was a brilliant researcher who got there first, this was a discovery whose time had come; given the technology of the 1920s it was inevitable that the discovery would be made sooner rather than later. In 1929, having carried out a similar analysis using a different technique, Russell himself published a paper confirming these results and giving due credit to Payne's priority. Unfortunately, because of Russell's established position in the astronomical community, for some time he was often cited as the discoverer by people who should have known better (or at least, should have read his paper properly).

Payne went on to a distinguished career in astronomy; in 1934 she married the Russian-born astrophysicist Sergei Gaposchkin and became known as Cecilia Payne-Gaposchkin. She remained at Harvard throughout her career, in spite of the low status and low pay she received as a woman. For many years, her official title was 'technical assistant', even though she carried out all the research and teaching duties expected of a professor. It was not until 1956 that she was promoted to become a full professor – the first female

---

* Much later, McCrea was my PhD examiner.

professor at Harvard. But, like most scientists, she was not primarily motivated by status or salary. In 1976, three years before her death, she was awarded the prestigious Henry Norris Russell Prize by the American Astronomical Society. No doubt she appreciated the irony. In her acceptance lecture, she said, clearly referring to her early work on stellar spectra: 'The reward of the young scientist is the emotional thrill of being the first person in the history of the world to see something or to understand something.' Even if someone else tells you it is 'clearly impossible'.

Even at the end of the 1920s, however, astrophysicists had yet to grasp the full significance of the discovery that the Sun's atmosphere is overwhelmingly rich in hydrogen. It would be nearly two decades before they would appreciate that even the *interior* of a star like the Sun is largely made of hydrogen (and some helium, but very little in the way of heavier elements). The longevity of this misconception was partly the result of an unfortunate coincidence, discussed later, which stemmed from the developing understanding of how hot stars are.

The heat of the Sun

Two temperatures are particularly important in our understanding of the nature of stars. The first is the temperature at the surface of the Sun; the second is the temperature at the heart of the Sun. They can be put in perspective with some basic solar vital statistics.

The key measurement is the distance from the Earth to the Sun. From the laws of planetary motion worked out by Johannes Kepler in the 17th century, we know that the distance from the Sun to Venus is 0.72 times the distance from the Sun to the Earth, but we do not know the exact distances. Fortunately, on rare occasions (most recently in 2012) Venus passes directly across the face of the Sun, as viewed from Earth. These 'transits of Venus', combined with Kepler's laws, make it possible to work out the Sun–Earth distance from parallax. If a transit is observed from two widely separated points on the Earth's surface, the moment when Venus crosses the edge of the Sun's disc will be seen at different times for the two observers, because they are looking at different angles. It is then easy to work out from simple geometry that the Sun is just under 150 million kilometres away from us. In order to look as big as it does at such a distance, it follows that the diameter of the Sun is about 108 times the diameter of the Earth.

We can also work out how massive the Sun is. The amount of matter in the Sun determines its gravitational pull, which holds the planets, including the Earth, in their orbits around the Sun. The earth orbits the Sun once every year at a distance of 150 million kilometres, so astronomers know how fast it is moving in its orbit. The force required to hold the planet in this orbit is known from basic physics, and it doesn't matter whether that force is provided by a long piece of string fastened at the centre of the Solar System or by the gravitational attraction between the Sun and the Earth. Knowing the force, we can use Newton's law of gravity to calculate that the mass of the Sun is about a third of a million times the mass of the Earth.* Since the volume of the Sun (proportional to diameter cubed) is a bit more than a million times the volume of the Earth, this means that the *average* density of the Sun must be a third that of the Earth, only 1.5 times the density of water. But as we shall see, that average is far from being the whole story.

So we know how far away the Sun is and how big it is. How hot is it? There are two ways to tackle this question. The first goes back to an observation made by William Herschel in the 18th century. He pointed out that the heat of the noonday Sun at the equator is sufficient to melt a layer of ice one-inch thick on the surface of the Earth

---

* The mass of the Earth has been known since the end of the 18th century, when the English physicist Henry Cavendish measured the strength of gravity in a series of very sensitive experiments.

in two hours and twelve minutes. Because the Sun is pouring out energy evenly in all directions, this means that there is enough heat to melt a whole spherical shell of ice one-inch thick, surrounding the Sun at the distance of the Earth, in that time – a shell one-inch thick and 300 million km in diameter. Now imagine this shell of ice shrinking in towards the Sun, getting smaller in diameter but always containing the same volume of ice, so that it gets thicker the closer it gets to the Sun's surface. At the Sun's surface, it would be more than a mile thick, but it would still thaw just as quickly. The temperature at the solar surface needed to do this is just under 6,000 K.*

This neat physical example is fine for the Sun, but no use in trying to work out the temperature of the stars. Fortunately, there is another technique that can be applied more generally, and that gives the same answer for the temperature at the surface of the Sun, so we know that it works. This stems from another piece of work by the prolific (and versatile) Gustav Kirchhoff.

The heat of the stars

In 1859, Kirchhoff's investigation of the radiation emitted by hot objects led him to formulate a rule sometimes known as Kirchhoff's law. Confusingly, there is another 'Kirchhoff's law', developed by the same man, which applies to the behaviour of electrical circuits. But the one that is relevant here can be paraphrased as: at any specified temperature the rate at which an object emits electromagnetic energy (heat and light) is the same as the rate at which it absorbs electromagnetic energy with the same wavelength (or frequency). In 1859, this was an inspired guess, but in 1861 Kirchhoff carried out experiments that proved him right, and in 1862 he introduced the idea of a 'perfect' emitter and absorber of radiation, called a

---

* Remember that physicists measure temperature in degrees Kelvin, or K, from the absolute zero of temperature, –273 °C. So 0 °C is 273 K, and so on.

black body. Such an object would absorb all the radiation that falls on it, but in return as it got hot it would radiate energy across the electromagnetic spectrum. Crucially, though, it would not radiate this 'black-body radiation' evenly at all wavelengths.

There's a very simple experimental setup that makes it possible to study black-body radiation in the laboratory. Take a metal box, or a sealed tin can, and make a tiny hole in it. Any radiation getting in through the hole bounces around inside the box and warms the walls. It is a perfect absorber of radiation and, therefore, as Kirchhoff showed, a perfect emitter of radiation. 'Perfect' here means that the radiation emitted by a black body does not depend on what the black body is made of, its size, shape or any other physical properties. All that matters is its temperature. As it gets hot, some of the radiation inside leaks out through the hole and can be studied using prisms, spectroscopes and so on. You can even heat it more forcefully using, say, a Bunsen burner. It doesn't matter how the box gets hot; the radiation is always the same. This is black-body radiation or, to give it a semi-obsolete but descriptive name, 'cavity radiation'. A key point to take away from this is that a 'black body' is not necessarily black. It may be a strong emitter of light and heat. Indeed, on this basis the Sun is an almost perfect black body – and so are the stars.

This is the key to taking their temperatures. In 1879, following up experiments carried out by John Tyndall in England, the Slovenian physicist Josef Stefan measured the total amount of electromagnetic energy emitted by objects at different temperatures. He derived an equation relating the amount of energy being radiated to the temperature, and used this to calculate the temperature at the surface of the Sun, finding it to be just under 6,000 K. The relationship found by Stefan was refined by Ludwig Boltzmann in 1884, who showed that it only applies accurately to black bodies; it is known today as the Stefan-Boltzmann law.

In 1893, Wilhelm Wien, working at the University of Berlin, rounded off this phase of the investigation of black-body radiation. A graph of the amount of energy radiated by a black body at different wavelengths rises smoothly from lower energies at shorter wavelengths to a peak at some intermediate wavelength, then slides down smoothly again to lower energies at shorter wavelengths. The position of the peak moves towards shorter wavelengths at higher temperatures, and Wien found that the temperature of the black body can be found simply by dividing the wavelength of the peak emission (measured in micrometres) into the number 2.898. This is Wien's law. For example, if the peak is at 4 micrometres (or 0.004 mm), the temperature of the black body is 724.5 K. Although it is very straightforward and easy to use, Wien's law is one of the most important tools of astrophysics. It means that astronomers can measure the temperatures of the surfaces of stars simply by measuring the peak wavelength of their emission. And this matches up with our everyday experience.

We all know that objects change colour when they get hot. In the days of open coal fires in houses, this was even more obvious. My father used to light his cigarettes from an iron poker thrust into the fire. The cold poker (at the same temperature as the room) was, of course, black. As it got hotter, it became red – red hot, ideal for lighting a ciggie. If he forgot to pull the poker from the fire in time, it would get hotter still and glow white hot. Although I never saw it put to the test, I imagine that if he had ever left it longer still it would have melted. Wien's law puts this in numerical terms. Spectroscopy can measure just how hot 'red hot' is, just how hot 'white hot' is, and subtle gradations from the dimmest red to the brightest blue (and, indeed, beyond the visible spectrum into the infrared and the ultraviolet). Stars come in different colours, and red stars are cooler than blue stars. Wien's law tells us the actual surface temperatures of the stars. They all lie, in round terms, in the range from about

3,000 K to about 30,000 K; so on this basis the Sun is a pretty ordinary star with a modest surface temperature. But this is only half the story. What is the temperature *inside* the Sun and other stars?

The heat inside

It happens that the temperature at the heart of a stable star depends only on the mass of the star, its brightness (which is related to its temperature) and its composition. It doesn't matter *how* the heart of a star is kept hot; it just has to be a certain temperature to provide the pressure to hold itself up against the inward pull of gravity. We know the mass of the Sun from its gravitational influence on the orbits of the planets, and once it was known that our local star is mostly composed of hydrogen and helium it was straightforward to work out that the temperature in the middle of the Sun is about 15 million K. If the Sun is an ordinary star, the temperatures inside other stars ought to be similar. But to prove this, astronomers needed to know the masses of at least some other stars. Fortunately, they were able to do so by applying the same gravitational rules that control the orbits of the planets around the Sun to star systems in which two stars are orbiting around each other (binary stars), and even triple systems. In fact, roughly half the stars we see in the sky are actually members of binary systems. Once again, spectroscopy provides a key to unlocking the numbers.

As Bunsen and Kirchhoff discovered, each element produces lines at distinctive wavelengths in the spectrum. But if the object that is producing the spectrum is moving relative to the instruments doing the measuring, the observed wavelengths of these lines get shifted. For an object moving towards us, the waves get squashed to shorter wavelengths (higher frequencies), and since blue light has shorter wavelengths than red light this is called a blueshift. For an object moving away from us, the waves get stretched to longer wavelengths (lower frequencies), and since red light has longer wavelengths than blue light this is called a redshift.* For objects moving at an angle to us, the situation is more complicated, but it can be unravelled with patience and skill. These shifts are known as Doppler shifts, in honour of the German physicist Christian Doppler, who studied the effect on sound waves in the 1840s. The important point is that these Doppler shifts depend on the speed with which the object producing the spectrum is moving, so it is possible to monitor the speed of binary stars as they move in their orbits around one another.

Astronomers knew from basic physics that there is only a fairly limited range of possible masses for bright stars. A ball of gas that has less than about one-tenth of the mass of our Sun will never get very hot inside, and will form a cool object rather like a larger version of the planet Jupiter, known as a brown dwarf. At the other extreme, any ball of gas with more than a few hundred times the mass of our Sun will get so hot inside as it tries to hold itself up against the pull of its own gravity that it will blow itself apart. In very round numbers (spelled out in the 1920s by the pioneering astrophysicist Arthur Eddington, who first inspired Cecilia Payne to take up astronomy), bright stars must have masses in the range from 0.1 to 100 solar masses. Happily for basic physics (and basic physicists), studies of real stars in binary systems confirm this. But they also show something even more important. There is a simple relationship between the mass of a star and its intrinsic brightness, or luminosity, and this points to the fact that stars with very different masses and luminosities all have roughly the same central temperature.

The qualification 'intrinsic' brightness is important. Stars that actually have the same brightness may look fainter or brighter than one another depending on their distance from us. Similarly, a star

---

* There is another way to produce a redshift, which is important in the cosmological discussion of Part Two of this book but is not relevant here.

that looks bright in the sky may actually be relatively dim but very close, while a star that looks faint may be extremely bright but far away. Since there are ways to find the distances to the stars (which I shall go into in Chapter Five), these effects can be unravelled and converted into the 'absolute magnitude', which is the brightness a star would have if viewed from a distance of 10 parsecs (about 32.5 light years).

The exact form of the mass–luminosity relation changes slightly across the whole range of stellar masses, but for masses in the range from 0.3 to 7 times the mass of our Sun the luminosity is proportional to the fourth power of the mass. So a star that is twice as massive as the Sun, for example, is sixteen times brighter than the Sun, because 2^4 is $2 \times 2 \times 2 \times 2 = 16$. In a related relationship, it turns out that the diameter of a star like the Sun is directly proportional to its mass; so this hypothetical two-solar-mass star will also be twice as big across as the Sun, not sixteen times as big. It was Arthur Eddington who realised what the mass–luminosity relation means – that all these stars have the same central temperature. We now know that this temperature is roughly 15 million K, but in the mid-1920s Eddington did not know that stars are chiefly composed of hydrogen and helium; Cecilia Payne's breakthrough had not yet been accepted. So he calculated a figure that is slightly too high, and, referring to the energy requirements of two specific stars by name, wrote in his book *The Internal Constitution of the Stars*, published in 1926, that:

> Taken at face value [this] suggests that whether a supply of 680 ergs per gram is needed (V Puppis) or whether a supply of 0.08 ergs per gram (Krueger 80) the star has to rise to 40,000,000° to get it. At this point it taps an unlimited supply.

Later in the book, he went into more detail. As a star forms from a collapsing cloud of gas, he said, it will:

contract until its central temperature reaches 40 million degrees when the main supply of energy is suddenly released [...] A star [then] must keep just enough of its material above the critical temperature to furnish the supply required.

The big question this raised in 1926 was: just what was the source of the energy required to keep a star like the Sun shining? Eddington thought he knew, and he would soon be proved right, opening the way not just to an understanding of the stars as they are today, but to a complete understanding of the whole life cycle of a star, and eventually to measurements of the ages of the oldest stars in the Universe.* But they had to crack the puzzle of the age of the Sun first.

---

*   Somewhat cavalierly, astrophysicists refer to the life cycle of an individual star as its 'evolution'.

2 0.008

At the heart of the Sun

By some measures, the Sun isn't very hot at all. I am fond of an example put forward by George Gamow in his book *A Star Called the Sun*, published in 1964. If a perfectly insulated coffee pot produced heat at the same rate per gram as the average heat produced per gram by the Sun, how long would it take for the pot to heat the water in the pot from room temperature to boiling point? The superficially surprising answer is that it would take many months for the coffee to boil. It takes 100 calories of energy to raise the temperature of 1 gram of water from 0°C to 100°C. But on average, each gram of the Sun's mass produces very little heat. The mass of the Sun is 2×10^{33} grams, and just under 9×10^{25} calories of heat cross its surface every second. So each gram of the Sun's mass produces just under 4.5×10^{-8} calories per second – less than half of one 10-millionth of a calorie per second. This is actually considerably less than the rate at which your body produces heat through the chemical processes involved in metabolism – but your blood doesn't boil, of course, because your body is not insulated and the heat escapes.

So the problem is not that the Sun is hot. Even burning coal could produce that much heat for a few – or a few thousand – seconds. The puzzle that confronted astrophysicists at the start of the 20th century was how a star like the Sun could stay hot for so long. The immense age of the Earth had become clear, during the 19th century, from the developing understanding of geology and evolution. These pointed to an age for the Earth and by implication a minimum age for the Sun, vastly longer than could be explained by any known process, such as the burning of a Sun-sized lump of coal.

A French connection

The first serious attempt to calculate the age of the Earth had been carried out in the 18th century by a French aristocrat, the Comte de Buffon. Buffon was immensely wealthy and devoted his life to science and public service; he died in 1788, just missing the turmoil of the French Revolution, but his son, who inherited the title, was guillotined in 1794. Among his many contributions to science, Buffon picked up on a remark made in the previous century by Isaac Newton in his *Principia*. Newton had mentioned that 'comets occasionally fall upon the Sun', and a notion had grown up among natural philosophers (as scientists were known then) that the Sun might be a glowing ball of iron, from which the Earth had been torn by the impact of a comet. Newton himself guessed, without doing any experiments or detailed calculations, that a ball of red-hot iron the size of the Earth would not cool down to the point where it became inhabitable for 'above 50,000 years'. This seems to have passed without comment, even though it implies an age for the Earth more than ten times longer than the age inferred by Biblical literalists.

Buffon took these ideas further by actually carrying out experiments to see how quickly differently-sized balls of red-hot iron cooled down. The experiments were not very sophisticated, but they were informative, up to a point. Buffon measured how long it took for the different balls of iron to cool from a glowing red heat to the point where they could just be touched without burning the skin. From these measurements, he extrapolated to work out how long it would take a red-hot ball of iron the size of the Earth to cool to the same state. According to legend, his assistants in this experiment were aristocratic ladies whose delicate hands were shielded by fine white gloves, through which they touched the iron. It turned out that Newton had not been that far off. Buffon estimated that the Earth would have taken over 75,000 years to cool to the point where life could emerge. Crude though it was, this was a genuinely scientific attempt at measuring the age of the Earth, published in the second half of the 18th century. But it was soon superseded by the work of a member of the next generation of great French scientists. The age he calculated was so enormous that, even in the early 19th century, he never published it, possibly for fear of a hostile reception from the Church, perhaps because he couldn't believe it.

Joseph Fourier served as a scientific adviser to Napoleon and held high office in the civil administration of France, being rewarded with the titles of Baron and then Comte.* He began his studies of the way heat flows through solid objects while prefect (governor) of the Department of Isère in Grenoble, in the first decade of the 19th century; his masterwork on heat flow appeared in 1822. Fourier carried out many experiments, such as heating one end of an iron bar and monitoring the way heat propagated to the other end, from which he derived equations to describe heat flow. He then applied those equations to the calculation of how long it would take an Earth-sized ball of molten material to cool. He made a crucial improvement on Buffon's thinking by realising that once the crust of the Earth solidified, it would act as an insulating blanket around the interior, greatly slowing the rate at which heat could escape. This is part of the reason why the core of the Earth is, we now know, still molten today. (Another reason is that heat is still being released in the core by radioactivity, which soon comes into the story of the Sun.) Fourier wrote down the equations that led to a

---

* In full, he was Jean-Baptiste Joseph Fourier, but usually known as Joseph.

number for the age of the Earth implied by taking all of this into account, and he must surely have worked out the answer to those equations for himself. But he never published it and never left even a scrap of paper with the number on. This estimate for the age of the Earth, and by implication the Sun, is not a few thousand years or a few tens of thousands of years, but 100 *million* years. But as the 19th century progressed, this number turned out to be too big to please the astronomers and too small to please the geologists and evolutionary biologists.

No free lunch

Around the middle of the 19th century, physicists developed an understanding of thermodynamics, the rules that govern the behaviour of hot objects and the way energy in the form of heat is transferred from one object to another – crucially, always from a hotter object to a cooler object, never the other way around, in any natural system. Part of the impetus for this work came from the development of steam engines as the power source for the Industrial Revolution; a better understanding of steam engines led to an improved theory of thermodynamics and a better understanding of thermodynamics led to improved designs of steam engines. One of the key features of this new science – arguably *the* key feature of 19th-century physics – was the proper scientific appreciation of the second law of thermodynamics, which has been called the most important law in science. This says, in everyday language, that everything wears out, you can't have something for nothing, and there ain't no such thing as a free lunch. Physicists realised that these laws apply to the Sun itself (indeed, to the whole Universe), and that therefore the Sun was not an eternal source of heat and light for the Earth. In 1852, the British physicist William Thomson, who put forward the second law in 1851 and was later ennobled as Lord Kelvin (so he is usually known by that name) wrote:

> Within a finite period of time past the earth must have been, and
> within a finite period of time to come the earth must again be,
> unfit for the habitation of man as at present constituted.

But how long was that 'finite period of time'? Several people puzzled over the problem, but the two who thought most deeply about
it and developed much the same idea were Kelvin in England and
Hermann von Helmholtz in Germany. They appreciated that gravity
was the most potent energy source known at the time, and Kelvin
followed up a suggestion, made in 1853 by John Waterston, that
the Sun might be kept hot by the energy released by a succession
of meteors striking its surface. Unfortunately, he soon found that
the energy released would be totally inadequate. Even gobbling up
whole planets wouldn't do the trick – if Mercury, the innermost
planet, fell into the Sun, it would only provide enough energy to
keep the Sun hot for seven years, and even Neptune, the most distant
large planet in the Solar System, could only provide enough energy
to keep the Sun going for a couple of thousand years.

Kelvin let the problem lie for the rest of the 1850s, during which
time Helmholtz offered a new variation on the gravitational theme.
He suggested, in 1854, that the whole Sun might be shrinking and
releasing gravitational energy in the form of heat as it did so.

Such a process is outside our everyday experience, but easy to
understand. Imagine a lump of rock the size of the Sun broken
into little pieces, then spread out with the pieces far apart from one
another, before they are allowed to fall together under the influence of gravity. As the pieces collide, they will release heat, just
as a meteorite striking the surface of the Earth releases heat. The
energy required to spread the pieces out into space is the same as
the energy that will be released when they fall back together, and
the same rules apply to atoms as to pieces of rock. So a collapsing
cloud of gas also converts gravitational energy into heat and gets hot

in the middle. The heat produces an outward pressure that pushes against the inward pull of gravity and slows the collapse. Helmholtz did not make a precise calculation of how much energy would be released by the collapse of a ball of gas the size of the Sun but merely pointed out that it would be a lot. Which left the way clear for Kelvin to return to the problem in 1860 and finish the job;* his results were published a couple of years later.

These calculations only showed how much total energy could be released by the collapse of a cloud of material with the mass of the Sun. Kelvin didn't worry, in the early 1860s, about how the energy might be, in a sense, stored up and released gradually over a long period of time. But he could still work out a maximum possible age for the Sun, simply by taking the total energy and dividing it by the rate at which the Sun is radiating energy today. The answer that he came up with was that, in round numbers, gravitational energy could keep the Sun shining at its present rate for 10 or 20 million years. Allowing a factor of ten for the possibility of errors in his calculations, in the published article describing this work he wrote:

> It seems, therefore, on the whole most probable that the sun has not illuminated the Earth for 100,000,000 years, and almost certain that he has not done so for 500,000,000 years. As for the future, we may say, with equal certainty, that the inhabitants of the Earth cannot continue to enjoy the light and heat essential to their life, for many million years longer, unless sources now unknown to us are prepared in the great storehouse of creation.[11]

But this statement appeared in print three years after the publication of *On the Origin of Species*. Charles Darwin, who was a geologist

---

* This was a year after the publication of Charles Darwin's *On the Origin of Species*; reading this book may have been what set Kelvin thinking about the timescale problem again.

before he became a famous biologist, was greatly influenced by the geologist Charles Lyell. Lyell's work on the age of the Earth – which explained how it had been shaped by processes such as volcanism, wind and weather – provided Darwin with the long timescale he needed for the process of evolution by natural selection to operate to produce the variety of life on Earth today. Darwin read Lyell's books on geology during his voyage on the *Beagle*. In a letter to a colleague, he wrote: 'I always feel as if my books came half out of Lyell's brains & that I never acknowledge this sufficiently.' Although, as this example shows, he always did acknowledge it. Influenced by Lyell, Darwin calculated how long it must have taken for the processes of erosion to have shaped the chalk hills and valleys of the English Weald, as an illustration of the long history of our planet. It was already clear, from the work of people like Lyell, that the Weald is a young feature, geologically speaking, so the Earth itself must be much older than the estimate Darwin came up with. It was a very rough calculation, and the number was rather too large, but not ludicrously so compared with modern understanding. Kelvin seized upon it with heavy sarcasm:

> What then are we to think of such geological estimates as 300,000,000 years for the 'denudation of the Weald'? Whether is it more probable that the physical conditions of the sun's matter differ 1,000 times more than dynamics compels us to suppose they differ from those of matter in our laboratories; or that a stormy sea, with possible channel tides of extreme violence, should encroach on a chalk cliff 1,000 times more rapidly than Mr Darwin's estimate of one inch per century?

Kelvin was only 38 in 1862, and for the rest of the 19th century he maintained and strengthened his arguments in favour of an age for the Earth and the Sun much less than the age required by geology and evolution. His not unreasonable point, from the state of knowledge at the time, was that there is no such thing as a free lunch, and that of all the forms of energy known to contemporary (19th-century) science, gravitational energy was the one that could supply the Sun with heat for longest. Following his calculation of the age of the Sun as a few score million years, Kelvin calculated the age of the Earth, assuming that it had formed as a ball of molten rock from the collision of meteors. He used Fourier's equations and calibrated the calculation with data from measurements of how the temperature inside the Earth rises as you go down deep mine shafts. His estimate of 98 million years for the age of the Earth was actually longer than his basic age for the Sun, but Kelvin was unfazed; it agreed nicely with the more cautious estimate that he was willing to publish. Applying the same caution, he said that the age of the Earth might be as little as 20 million years or as great as 200 million years, but no more. But as time passed, improved calculations pushed the numbers down towards the lower end of the range, just as the geologists and evolutionary biologists were pushing their estimates in the opposite direction.

The final refinement of his idea was presented by Kelvin in a lecture at the Royal Institution in London in 1887. It actually drew on a suggestion made by Helmholtz in his 1854 paper, but with numbers included. The resulting estimate for the age of the Sun (and stars) is usually known today as the Kelvin–Helmholtz timescale; it is based on the idea that the Sun is slowly shrinking under its own weight and steadily releasing gravitational energy in the form of heat as it does so.

This is the picture I mentioned earlier, of a cloud of gas in space collapsing down under its own weight and getting hot inside as gravitational energy is converted into kinetic energy of the atoms bashing against each other in the core and getting hotter. By the time such a collapsing cloud has shrunk to the size of the Sun, the temperature in the core will be millions of degrees – providing the pressure that opposes the inward tug, gravity – and the surface will have a temperature of a few thousand degrees. This is exactly how astronomers now think stars form, collapsing and settling down on the Kelvin–Helmholtz timescale.

But once the proto-star is hot inside, its collapse slows dramatically. As long as the star is hot inside, it can never collapse completely. If it cooled off, the pressure would drop and the star would shrink. But as it shrank, it would release gravitational energy and get hot, increasing the pressure and slowing the collapse. Kelvin was able to calculate how much the Sun would have to shrink each year in order to release the amount of energy measured as radiating from its surface. The answer is just 50 cm per year, or 50 metres per century. Shrinking at a rate of 50 metres per century, an amount far too small to be measured by 19th-century astronomers, the Sun could keep shining for 20 or 30 million years. But no longer.

Kelvin's dogmatism had not mellowed with age. In 1889, he wrote:

> It would, I think, be extremely rash to assume as probable anything more than twenty million years of the sun's light in the past history of the Earth, or to reckon on more than five or six million years to come.[12]

In 1897, the year he was elevated to the peerage, having settled on 24 million years as the best estimate for the age of the Sun and Earth, he reiterated:

> Within a finite period of time past the earth must have been, and within a finite period of time to come must again be, unfit for the habitation of man as at present constituted, unless operations have been and are to be performed which are impossible under

the laws governing the known operations going on at present in the material world.

By 'finite period' he meant 24 million years, and he intended the sentence as a put-down to the geologists and evolutionists. In fact, 'operations' going beyond the previously known laws of physics had just been discovered and would transform our understanding of the stars in the 20th century.

Seats of enormous energies

In 1899, the American geologist Thomas Chamberlain, responding to the timescale problem posed by astronomers, wrote in the journal *Science*:

> Is present knowledge relative to the behavior of matter under such extraordinary conditions as obtained in the interior of the sun sufficiently exhaustive to warrant the assertion that no unrecognised sources of heat reside there? What the internal constitution of the atoms may be is yet open to question. It is not impossible that they are complex organisations and seats of enormous energies. Certainly no careful chemist would affirm that the atoms are really elementary or that there may not be locked up in them energies of the first order of magnitude. No cautious chemist would [...] affirm or deny that the extraordinary conditions which reside at the center of the sun may not free a portion of this energy.

He was right. Indeed, the revolution that transformed astrophysics (and many other branches of science) had already begun, in 1895, with the discovery of X-rays.

The discovery came when Wilhelm Röntgen, already an eminent, 50-year-old professor at the University of Würzburg, was investigating the radiation emitted by a negatively-charged plate (the cathode, hence cathode rays) in an evacuated glass tube. We now know that these 'rays' are actually particles, which became known as electrons, but that wasn't discovered until 1897, by J.J. Thomson (no relation to Lord Kelvin). Röntgen found that where the cathode rays struck the wall of the glass tube they caused the emission of another kind of radiation, mysterious 'X-rays', which were soon shown to be a form of electromagnetic waves, like light but with much shorter wavelengths. Important though the discovery was, it did not fly in the face of the known laws of physics. Energy from the cathode rays was making the spot on the glass tube fluoresce, and this was converting some of the energy into X-rays. But the next development was much more startling.

Röntgen's discovery, announced on 1 January 1896, immediately triggered a wave of interest in fluorescence and raised the question of whether substances that fluoresce naturally under the influence of sunlight might produce X-rays, or something similar to them. One of the people who took up the challenge was Henri Becquerel in Paris. The distinctive feature of X-rays is, of course, that they can penetrate material such as cloth, paper and even flesh – as Röntgen had discovered. Among the many fluorescent materials he studied, Becquerel found some crystals (potassium uranyl disulphate) that glowed after being exposed to sunlight (fluoresced) and that emitted radiation that could fog a photographic plate even when the plate was wrapped in two layers of thick black paper.

Planning to investigate the phenomenon further, Becquerel prepared another double-wrapped photographic plate, put a copper cross on top of it, stood a dish containing the crystals on top of that, then put the whole setup in a cupboard, waiting for a sunny day so that he could get the crystals to fluoresce. But this was in late February 1896, when Paris remained overcast for several days. Eventually, Becquerel got bored of waiting, and on an impulse he developed the photographic plate anyway. To his astonishment, he saw the clear outline of the metal cross on it. Even without an input of energy from the Sun, without fluorescing, the crystals had produced radiation that travelled in straight lines and fogged the plate everywhere except where it had been shielded by the metal, which the radiation could not penetrate. The radiation was dubbed radioactivity and was soon found to be coming from uranium in the crystals used by Becquerel – even though pure uranium does not fluoresce. Later that year, Becquerel wrote in the journal *Comptes Rendus* that: 'one has not yet been able to recognise wherefrom uranium derives the energy which it emits with such persistence'. This was a much bigger puzzle than X-radiation, because the energy seemed to come from nowhere, violating one of the most basic principles of physics, that you can't get something for nothing. The energy of Röntgen's X-rays came from the electrons hitting the glass of the tube; the energy of fluorescence came from sunlight. But where did the energy in radioactivity come from?

Becquerel made his discovery serendipitously. The team who picked up the discovery and made an exhaustive study of radioactivity were Marie and Pierre Curie, also working in Paris. Working under extremely difficult (and, we now know, dangerous) conditions, the Curies identified and isolated two other, previously unknown, radioactive elements: polonium and radium. For this work, they shared the Nobel Prize with Becquerel in 1903. Their story has often been told, and there is no need to go into details here. The key point that is relevant to the ages of the Sun and stars relates to measurements made by Pierre Curie and his assistant Albert Laborde in the same year that the Curies received the Nobel Prize. They measured the amount of heat produced by a sample of radium sitting in isolation, with no input of energy from the outside world. It turned out that a single gram of pure radium (*every* single gram of pure radium) releases enough energy in one hour to raise the temperature of 1.3 grams of water from 0°C to 100°C, or to melt its own weight of ice. For a time, it looked as if the law of conservation of energy had been broken. Unable to believe that, Kelvin, now 79, asserted that energy must be being supplied to the radium from outside, that 'somehow ethereal waves may supply energy to radium'. Both hypotheses were wrong. The theoretical basis of an understanding of what was going on was, in fact, about to be supplied by a young technical assistant at the patent office in Bern. But before I introduce him, I should complete the story of the experimental investigation of radioactivity.

Ernest Rutherford, a New Zealander working in Cambridge, also measured the heat being released by radium in 1903 and went on to probe the structure of the atom. In the late 1890s Rutherford had been working as a research student in the same laboratory (the Cavendish) where J.J. Thomson revealed the particle nature of the electron. There, he helped to prove that X-rays are a form of electromagnetic wave, and went on to investigate the radiation discovered by Becquerel. He found that this radiation has two components, which he named alpha and beta rays. Alpha radiation has a short range and can be stopped by a piece of paper; beta radiation has a longer range and more penetrating power. He later identified a third kind of radioactive 'ray', which he called gamma radiation. Further investigations showed that alpha rays are made up of streams of particles identical to helium ions – helium atoms from which two electrons have been stripped. (This discovery came in 1908, the same year that Rutherford was awarded the Nobel Prize, and little more than ten years after helium was first found on Earth.) Beta rays are fast-moving electrons. Gamma rays are electromagnetic waves, like light and X-rays but with even shorter wavelengths.

Rutherford moved from Cambridge to McGill University in Montreal in 1898, then back to England to the University of Manchester in 1907. While in Canada, working with Frederick Soddy, Rutherford discovered that when a radioactive atom emits alpha or beta radiation, in the process now known as radioactive decay, the atom is transformed into an atom of a different element. When a radium atom, for example, emits an alpha particle, it is transformed into an atom of the gas radon. Rutherford received the Nobel Prize in Chemistry for this work, with the citation specifying his 'investigations into the disintegration of the elements and the chemistry of radioactive substances'. This was somewhat ironic, since Rutherford always looked down on chemistry and once said that 'all science is either physics or stamp collecting'.[13] But his most important work, for which he should have received the Physics Prize (but didn't) still lay ahead of him.

Rutherford and Soddy also found that the radioactivity associated with the disintegration of atoms could not provide an infinite source of energy. They showed that there is a characteristic timescale for this process, and that for each radioactive element, half of the atoms present in a sample will decay in a certain time, a time uniquely associated with that particular element, known as the half-life. In the next half-life, half of the remaining radioactive atoms (a quarter of the original) will decay, and so on. The half-life for radium is quite short, compared with cosmic timescales: just 1,602 years; however much you start with, as time passes the amount of radioactivity and the amount of heat being generated decreases.* This was a clue that the storehouse of energy present in radioactive materials had been built up by some unknown process long ago, and was now being depleted, rather like the way deposits of coal represent a finite storehouse of energy from the Sun absorbed and laid down by plants.

In Manchester, the year after receiving his Nobel Prize, Rutherford directed research by Hans Geiger and Ernest Marsden which used the newly-discovered alpha particles to probe the

---

* Radium is present on Earth today because it is one of the products of the decay of much longer-lived uranium atoms.

structure of matter. By directing beams of alpha particles from radio-active material at gold foil, they discovered that although most of the particles passed right through the foil, a few hit something solid and bounced back the way they had come. This led Rutherford to develop his model of the atom as a tiny central nucleus, containing almost all the mass of the atom and with a positive charge, surrounded by a cloud of negatively-charged electrons, through which alpha particles (themselves now identified as helium nuclei) could pass unhindered. Only on the rare occasions that they approached a nucleus head on were they deflected, as the positive charge on the nucleus repelled the positive charge on the alpha particles. This discovery was well worth a Nobel Prize.

While all these developments were going on, though, Rutherford had also found time to address the question of the source of the energy which keeps the Sun and stars shining. As early as 1899 Rutherford had commented that the origin of the energy in Becquerel radiation was 'a mystery', and in 1900, working at McGill with R.K. McClung, he had shown just how much energy was carried by the different kinds of rays emitted during radioactivity. Around the same time, two German schoolteachers, Julius Elster and Hans Geitel, showed that the source of energy had to lie within the radioactive material and could not come from outside. They put radioactive materials in vacuum jars down deep mine shafts, away from any source of external energy such as the Sun, and found no decrease in their activity. At the beginning of the 20th century, they showed that there is natural radioactivity all around us, at a low level, in the air and soil, and other researchers found radioactivity in the rocks. This led George Darwin (one of the sons of Charles Darwin) and John Joly to suggest that radioactivity might be at least partly responsible for the heat of the Sun, and Robert Strutt, of Imperial College, London, to propose that the presence of radioactive substances such as radium inside the Earth could provide the source of energy needed to explain the length of the geological timescale. This was before Rutherford and Soddy discovered the half-life rule, but Strutt was not far from the truth, because longer-lived radioactive substances do indeed provide some of the heat in the Earth's core today.

Rutherford maintained his interest in the problem over the next few years. Shortly after the measurements of heat production by radium carried out by Curie and Laborde, he was able to show, working with Howard Barnes, that the amount of heat produced by radioactivity depends on the number of alpha particles emitted. It was clear that the heat was generated by alpha particles from radioactive atoms colliding with other atoms (actually, as Rutherford would soon discover, with other nuclei) and jostling them, converting the kinetic energy of the alpha particles into the heat energy of their surroundings. Armed with this discovery, in 1904 Rutherford also suggested that radioactive decay might solve the puzzle of the age of the Earth. He presented the idea at a meeting of the Royal Institution in London, where Kelvin, now a venerable elder statesman of science, was present:

> I came into the room, which was half dark, and presently spotted Lord Kelvin in the audience and realised that I was in for trouble at the last part of the speech dealing with the age of the earth, where my views conflicted with his [... A] sudden inspiration came, and I said Lord Kelvin had limited the age of the earth, *provided no new source of heat was discovered*. That prophetic utterance refers to what we are now considering tonight, radium! Behold! The old boy beamed upon me.[14]

Although Rutherford naturally plays up the importance of his own contribution to the debate, the idea that radium might provide energy to keep the Sun hot was becoming widespread by 1904. Following the work by Curie and Laborde, the English astronomer William Wilson published a paper in *Nature* in July 1903, in which he showed that just 3.6 grams of radium in every cubic metre of the Sun's substance would be enough to supply all of the heat being radiated today – although, again, he was unaware at the time of the half-life problem. It was this idea that was taken up by George Darwin, also writing in *Nature*, who rather cautiously suggested that Lord Kelvin's timescale for the age of the Sun might be extended by a factor of ten or twenty, out to about a billion years. But the main objection to this idea was that spectroscopic studies showed no trace of radioactive elements, such as uranium and radium, in the Sun. In 1905, though, a possible ultimate source of the energy in radioactivity was discovered.

The discoverer was, of course, Albert Einstein, with his special theory of relativity. The famous equation $E = mc^2$ did not appear in the paper that introduced the special theory of relativity to the world. That paper was titled 'On the Electrodynamics of Moving Bodies' and published in the journal *Annalen der Physik* at the end of September 1905. But in the same week that this paper was published, the editor of that journal received another paper from Einstein, just three pages long, which would also be published before the end of 1905. In this he spelled out the implication from the special theory that matter is a form of stored up energy, and that mass and energy are interconvertible – although he used the letter L for energy and V for the speed of light; so even in this paper the famous equation did not appear in its now-familiar form. Einstein's thinking, including his awareness of the implications for an understanding of radioactivity, is spelled out clearly in a letter he had written in the summer of 1905, to Conrad Habicht:

> One more consequence of the paper on electrodynamics has also occurred to me. The principle of relativity, in conjunction with Maxwell's equations, requires that mass be a direct measure of the

energy contained in a body; light carries mass with it. A noticeable decrease in mass should occur in the case of radium.

A hotter place?

A gradual decrease in mass of the Sun could therefore explain the origin of the energy that it was pouring out into space. Using Einstein's equation, it is simple to calculate that the loss of mass implied for the Sun is about 4 million tons every second. This is mind-bogglingly large by human standards, but the Sun is itself so mind-bogglingly large that if it kept doing this for a trillion years it would only have used up less than one hundredth of its mass. If you believed Einstein (and not everyone did, at first), the timescale problem of geology and evolution was solved. But how was the Sun converting mass into energy?

On this occasion, theory had raced ahead of experiment, and more data would be needed before progress could be made in understanding what was going on inside the Sun and stars. The key experimental discovery was made by Francis Aston, working at the Cavendish Laboratory in Cambridge, in 1919. He developed an instrument called a mass spectrograph, or mass spectrometer, which could be used to measure the masses of atoms of a chosen element. It works by first ionising the atoms, then bending a beam of the resulting ions with a magnetic field. For a particular strength of magnetic field, the amount by which the beam is bent depends on the mass of the ions. The fact that the instrument uses a beam of ions rather than individual ions does not matter, since all the ions with the same mass are bent by the same amount, so the bending of the whole beam does reveal the mass of the individual atoms. Aston received the Nobel Prize for his work, in 1922. But one of the first discoveries he made with his new instrument was that an atom of helium has a mass 0.008 (that is, 0.8 per cent) less than the mass of four atoms of hydrogen put together. Other atomic masses were also found to be nearly, but not quite, exact multiples of the mass of a hydrogen atom, refining estimates made from chemical studies. So it was widely accepted that the other elements must in some sense be built up from hydrogen. This idea was powerfully reinforced in 1919 when Rutherford was able to convert nuclei of nitrogen into nuclei of oxygen by bombarding a nitrogen 'target' with alpha particles – transmutation of one element into another.

Arthur Eddington, fresh from his triumphant confirmation of the general theory of relativity, seized on the implication of all this from the perspective of the special theory. Addressing a meeting of the British Association for the Advancement of Science, held in Cardiff in August 1920, he made one of the most prescient statements in the history of astronomy:*

Only the inertia of tradition keeps the contraction hypothesis alive – or rather, not alive, but an unburied corpse. But if we decide to inter the corpse, let us freely recognise the position in which we are left. A star is drawing on some vast reservoir of energy by means unknown to us. This reservoir can scarcely be other than the sub-atomic energy which, it is known, exists abundantly in all matter; we sometimes dream that man will one day learn to release it and use it for his service. The store is well-nigh inexhaustible, if only it could be tapped. There is sufficient in the Sun to maintain its output of heat for 15 billion years [...]

Aston has further shown conclusively that the mass of the helium atom is even less than the masses of the four hydrogen atoms which enter into it – and in this, at any rate, the chemists agree with him. There is a loss of mass in the synthesis amounting to 1 part in 120, the atomic weight of hydrogen being 1.008 and that of helium just 4. I will not dwell on his beautiful proof

---

* Note that the atomic weight of helium is defined as four, in his example, and other atomic weights measured relative to that.

of this, as you will no doubt be able hear it from himself. Now mass cannot be annihilated, and this deficit can only represent the mass of the electrical energy set free in the transmutation. We can therefore at once calculate the quantity of energy liberated when helium is made out of hydrogen. If 5 percent of a star's mass consists initially of hydrogen atoms, which are gradually being combined to form more complex elements, the total heat liberated will more than suffice for our demands, and we need look no further for the source of a star's energy.

If, indeed, the sub-atomic energy in stars is being freely used to maintain their great furnaces, it seems to bring a little nearer to fulfilment our dream of controlling this latent power for the well-being of the human race – or for its suicide.*

This was still, of course, half a dozen years before Cecilia Payne discovered that the Sun and stars are *mostly* made of hydrogen, and nearly ten years before that idea was fully accepted. But apart from that, Eddington hit the nail on the head. There was, though, one snag.

By the middle of the 1920s, when Eddington was writing his book on *The Internal Constitution of the Stars*, although it was clear that the conversion of hydrogen into helium could indeed, in principle, provide ample energy for the needs of the Sun and stars, the problem was that calculations based on theory and the results of experiments like those in which nitrogen is transmuted into oxygen said that, even with a temperature of tens of millions of degrees, the centre of the Sun was not hot enough for the transmutation of hydrogen into helium to take place.

A simple way to understand the problem is in terms of the electrical repulsion between two positively-charged particles. Hydrogen

---

* Eddington spoke as a Quaker, with the horrors of the First World War fresh in his mind.

nuclei consist of single positively-charged protons, and when they approach one another, they are repelled from each other by the effect of these charges. Crudely speaking, in order for the transmutation (nuclear fusion) to take place, the protons have to collide physically, touching one another. Once this happens, they can stick together because of a short-range attractive force (poorly understood in the 1920s) which overpowers the electrical force; this is known as the strong nuclear force. The hotter they are, the faster the protons will move, and the faster they move, the closer together they will get. But the physicists told the astronomers that conditions in the heart of the Sun were not extreme enough for protons to get sufficiently close to each other for fusion. Eddington rejected their arguments. He had faith in the simple laws of physics that he had used in calculating the temperature at the heart of the Sun, and he was convinced that fusion of hydrogen into helium was the only way to explain how stars kept shining for so long. So in his book he wrote that: 'The helium which we handle must have been put together at some time and some place.' He dismissed his critics with: 'We do not argue with the critic who urges that the stars are not hot enough for this process; we tell him to go and find *a hotter place.*' This was interpreted as his way of telling his critics to go to hell.

Eddington was both right and wrong. Right that helium is manufactured out of hydrogen inside the Sun, releasing energy in line with Einstein's equation; wrong in thinking that all the helium in the Universe has been made inside stars in this way. But it is the right part of his argument that matters here. Astrophysics was rescued from the dilemma by dramatic developments going on in another branch of physics exactly at the time Eddington was writing those words. As he wrote in a Preface, dated July 1926: 'as we go to press a "new quantum theory" is arising which may have important reactions on the stellar problem when it is more fully developed.' In this, he was 100 per cent correct.

A quantum of solace

Quantum theory was born out of the investigation of black-body radiation, which also proved crucial (as we have seen) in understanding the nature of stars and (as we shall see) in understanding the Universe. It began with the work of the German physicist Max Planck, right at the end of the 19th century. He showed that the shape of the curve of a black-body spectrum could only be explained if atoms were only able to emit or absorb electromagnetic radiation, including light, in discrete chunks.* Planck knew perfectly well that light behaved like a wave, so did not suggest that it only existed in chunks, or as a stream of particles, but he did suggest that something in the nature of atoms made it impossible for them to interact with these waves except in finite energy steps. It was Einstein, in 1905, who suggested that these units of electromagnetic energy might be real, particles that are now known as photons. This is the work for which he received the Nobel Prize. He refined the understanding of the particle nature of light with further work in the 1910s, and in the 1920s in collaboration with Satyendra Bose.

So, in the mid-1920s there was clear evidence that light behaved as a wave (not least from experiments in which the light waves were made to interfere with each other, like ripples on a pond, to make a diffraction pattern). There was also clear evidence that light was made up of particles (not least from experiments in which electrons were knocked out of metal surfaces by photons). But in 1924, the Frenchman Louis de Broglie had put forward the idea (spelled out mathematically and endorsed by Einstein) that if electromagnetic 'waves' must also have a particle character, then all material 'particles' such as electrons must also have a wave character. This was soon confirmed in separate experiments carried out in England

---

* This simple statement conceals a vast effort by Planck; see my book *In Search of Schrödinger's Cat.*

by George Thomson (the son of J.J.) and in the USA by Clinton Davisson and Lester Germer. De Broglie, Davisson and Thomson all received Nobel Prizes as a result; Germer, as a research student and regarded as Davisson's 'assistant', missed out. A clear indication of the nature of quantum reality is that J.J. Thomson got the Nobel Prize for proving that electrons are particles, while his son got the Nobel Prize for proving that electrons are waves, and both were right.

So by 1926, when Eddington's book was published, it was becoming apparent that all quantum entities have both wave-like and particle-like aspects to their nature. The waves are mostly confined within a small volume, as a wave 'packet'; but this is still much bigger than the image of a point-like particle, such as an electron, and gives a certain fuzziness even to objects, such as alpha particles, that were initially thought of as little spheres. This is related to the famous uncertainty principle of Werner Heisenberg, although it would be wandering too far from my present astrophysical tale to go into detail. What matters here is that by 1928 a young Russian physicist, George Gamow, had applied these ideas to solve a major puzzle in nuclear physics.

The puzzle Gamow solved seems, at first sight, to be the opposite of the puzzle Eddington was confronted with in 1926. How could particles escape from a nucleus in the radiation process known as alpha decay? It's all to do with the balance between the attractive strong nuclear force and the repulsive electrical force. Together, they combine to create what is known as a potential well, which you can think of as like the crater of an extinct volcano. You can imagine alpha particles and the other particles that make up an atomic nucleus as like balls rolling about inside the crater. If one of the balls (alpha particles) is rolling fast enough (if it has enough energy) it can roll up the inner wall of the crater and over the top. Then, it will roll away down the other side and away. It has enough energy to escape from the volcanic crater, and no matter how slowly it is going when it goes over the top, once it has overcome the strong force it is pushed away by the electric force.

But in the mid-1920s, all the evidence from theory and experiment was that, according to the classical laws of physics (that is, the laws worked out in the pre-quantum era), alpha particles inside a nucleus did not have enough energy to escape. It was Gamow who realised how the quantum rules changed this conclusion. He pointed out that the position of a wave packet is uncertain and does not have a hard edge like a ball. When the alpha 'particle' is near the top of the crater rim, where the wall is thinnest, the wave may actually be large enough to stick out on the other side of the rim and feel the electric force of repulsion. This can drag the whole wave – the whole particle – through the wall, a process which became known as the 'tunnel effect'. The quantum rules made it possible to calculate how likely it was that this tunnelling would occur in different nuclei, matching up with the experimental observations.

This was like the cartoon image of a light bulb lighting up above the heads of the physicists. If alpha particles could tunnel *out* of a nucleus in this way, even when classical theory said that they did not have enough energy to do so, then maybe protons could tunnel *in* to nuclei and build up there to make helium nuclei, released as alpha particles and energy, even when classical theory said that they did not have enough energy to do so at the temperatures known to exist at the heart of the Sun and stars. It would be like two wave packets getting close enough for their tips to overlap, feel the strong force and tug each other together in a mutual embrace. All that remained was to work out the details of the process. But that was easier said than done. Gamow's idea was published in 1928, before Payne's work was widely appreciated, and at first astrophysicists trying to solve the puzzle were handicapped by the idea that stars are mostly made of elements much heavier than hydrogen.

3 7.65

Making 'metals'

In 1928, the best guess that physicists could make about the composition of a nucleus of helium – an alpha particle – was that it consisted of four protons and two electrons, bound together by the strong force. The four protons were needed to explain the mass of the alpha particle, but on their own that would have given it a positive charge of four units, two more than the actual charge of an alpha particle. So the two lightweight but negatively-charged electrons were needed to balance the electrical books. It was only in 1932 that James Chadwick, working in the Cavendish laboratory, discovered the uncharged particles now known as neutrons, which have slightly more mass than protons. This immediately made it clear that helium nuclei are each made up of two protons and two neutrons, held together by the strong force, with each helium atom completed by the addition of two electrons far out from the nucleus, held in place by electrical forces constrained by the rules of quantum physics. But the first steps towards an understanding of the processes that combine protons together to make helium and heavier elements – nuclear fusion – were made even before Chadwick's breakthrough.

Gamow's discovery of the tunnel effect was the direct inspiration for the first step, made by the physicists Robert Atkinson and Fritz Houtermans, and published in 1929. In their paper, they wrote: 'Recently Gamow demonstrated that positively-charged particles can penetrate the atomic nucleus even if traditional belief holds their energy to be inadequate.' They went on to describe mathematically how a heavy nucleus might absorb four protons* in this way, one at a time, and then spit out a complete alpha particle. Their mistake – if you can call it that – was to think that the composition of the Sun was similar to the composition of the Earth: that there would be plenty of heavy nuclei around in which this process could take place. They did not realise, nor did anyone else at the time, that the key is the direct interaction of protons with one another. But this gap in their knowledge is far less important than the fact that they put numbers into their calculations, which makes it possible to work out how many nuclear interactions would be needed each second to keep the Sun shining. The answer is surprisingly small, which makes the potential age of a star like the Sun correspondingly large.

Updating their idea, we can calculate that even under the conditions which exist at the heart of the Sun (where modern estimates say that the temperature is about 15 million K), only the fastest protons are able to tunnel through the electrical barrier. For any temperature, the particles in a fluid such as the material the Sun is made of are moving at different speeds, but the average speed is greater for higher temperatures. The speeds of individual particles are distributed around this average, with some moving faster and some slower, in accordance with well-known statistical laws. So it is possible to calculate what proportion of the distribution are moving, say, 10 per cent faster than the average, or 20 per cent faster, twice as fast as the average, and so on. The fastest moving particles are said to be in the 'high speed tail' of the distribution.

Updating the kind of calculations made by Atkinson and Houtermans reveals just how little nuclear fusion is needed to keep the Sun shining. In order for two protons to fuse with one another inside the Sun, they have to collide almost exactly head on, and one of them must be moving at least five times faster than the average

---

* And two electrons, in their model, which pre-dated the discovery of neutrons.

speed, way out in the high speed tail. Just one proton in every 100 million is moving fast enough to do the trick, and one collision in every 10 trillion trillion (1 in10^{25}) results in fusion.[15] On average, an individual proton can bounce around inside the Sun, involved in one collision after another like the ball in a crazy cosmic pinball machine, for 14 billion years before it fuses with another proton, and then participates in further interactions to make helium. Nuclear fusion is an extremely rare process, even at the heart of the Sun. But there are so many protons at the heart of the Sun that each second 616 million tons of hydrogen nuclei (protons) are converted into 611 tons of helium nuclei (alpha particles), with 5 million tons of mass converted into energy, in line with Einstein's equation. And yet, there is so much hydrogen in the Sun that it would take 5 billion years to convert 4 per cent of the original material into helium in this way. The timescale 'problem' of geology and evolution disappears at a stroke.

It was Atkinson (after Houtermans had moved on to other work) who showed, in the 1930s, that fusion of two protons to make a nucleus of deuterium (a deuteron), consisting of a single proton and a single neutron bound together by the strong force, is indeed the most likely first step in the manufacture of helium, the source of the Sun's energy. He started out with the idea that heavier nuclei were involved, but by 1936 it was clear that the Sun contained a great deal of hydrogen, and equally clear that the proton–proton interaction was the key step in nuclear fusion inside the Sun. It is easy to understand why. Heavier nuclei contain more protons so they have larger positive charge, and therefore electric repulsion makes it harder for incoming protons to tunnel into them. As it turned out, heavier nuclei are indeed involved in the way Atkinson and Houtermans originally suggested in some other stars, where conditions are more extreme. But even in 1936 there was still confusion about just how much hydrogen the Sun contained.

This confusion was related to an unfortunate coincidence which sent astrophysicists up a blind alley in the early 1930s. The kind of calculations pioneered by Arthur Eddington, which described the basic structure of a star like the Sun in terms of the physics of a ball of hot material, and told astronomers how hot it must be at the centre, depend on the composition of the star. There is a balance between gravity, pulling everything together, and pressure, including the pressure of electromagnetic radiation (light and other wavelengths), trying to blow the star apart. Radiation pressure is important because electromagnetic radiation interacts strongly with charged particles, and there are a lot of charged particles – negative electrons and positive nuclei – inside a star. If there were too many charged particles, they would hold the radiation in, and the star would swell up. Too few, and the radiation would escape, letting the star deflate like a pricked balloon. In fact, this is more or less what does happen when a star forms. If it shrinks, it gets hotter inside, producing more radiation which stops it shrinking; if it expands, it cools inside, producing less radiation which stops it expanding. But what Eddington and his contemporaries were interested in was the equilibrium situation when everything is in balance.

This equilibrium is affected by another factor – not just the number of charged particles, but how they are arranged. A nucleus of the most common form of iron, for example, contains 26 protons and 30 neutrons packed together. If all the protons in a star were packed into iron nuclei, the balance with radiation would be different from the balance that would be struck if all the protons were roaming freely, because either way, there is one electron around (roaming freely and able to interact with electromagnetic radiation) for each proton.

The critical factor, which could only be properly taken into account once neutrons had been discovered, is the number of electrons per nucleon, where 'nucleon' is a generic name for protons and neutrons. If a star were completely composed of hydrogen, this factor would be one, since all the nucleons would be protons, and there would be one electron for every proton. If the star were entirely made of helium, the number of electrons per nucleon would be 0.5, since there are four nucleons in a helium nucleus, but only two of them are positively-charged protons, so only two electrons are needed to maintain the electrical balance. If a star were entirely composed of iron, the number of electrons per nucleon would be 20 divided by 56, or roughly 0.35. Once astrophysicists realised that there is a great deal of hydrogen inside the Sun, they revised the kind of calculations that Eddington had made to allow for this.

But they found a curious thing. The calculations showed that a globe of material the size of the Sun, with all the observed external properties of the Sun, such as its surface temperature, could exist in either of two stable states. It could be in equilibrium if 35 per cent of the material in its interior was in the form of hydrogen. Equally, though, it could form a stable star if at least 95 per cent of its material was in the form of either hydrogen or helium, with just a smattering of every other element. Having previously thought that the composition of the Sun was more or less the same as that of the Earth, astrophysicists were now forced to accept that at least a third of it must be in the form of hydrogen. That was, initially, as far as they could bring themselves to go; the notion that 95 per cent of the Sun (and, by implication, the stars) was hydrogen or helium was too big a jump. This misconception – for that is what it was – coloured their thinking for decades, into the 1950s. But it didn't stop them from finding out exactly how stars do release energy by converting hydrogen into helium, and going on to make the first accurate estimates of the ages of stars.

Cycles and chains of fusion

George Gamow comes back into the story here. In 1938 he organised a conference in Washington DC, where astronomers and physicists got together to discuss the problem of energy generation inside stars. One of the participants at the meeting was Hans Bethe, a 31-year-old German physicist who, like so many of his colleagues, had moved to America after the rise of Hitler. The key problem addressed at the meeting was the question of exactly which nuclear fusion processes could generate the right amount of heat, at the temperature known to exist inside the Sun, to maintain the steady flow of energy from the Sun. By 1938, there was a great deal of experimental evidence to draw on, which told the physicists how quickly different reactions took place. At one extreme, for example, if there were a lot of lithium in the heart of the Sun, this would be rapidly converted into helium by interactions with hydrogen nuclei, generating so much energy that it would blow the Sun apart. At the other extreme, if the Sun were largely made of oxygen and hydrogen, the reaction between oxygen nuclei and protons would take place so slowly that the star would collapse, shrinking until it got hot enough inside to stimulate more vigorous interactions. The puzzle was to find a set of interactions that would be, like Baby Bear's porridge in the Goldilocks story, 'just right'.

Nobody solved the puzzle at the meeting, but in his book *The Birth and Death of the Sun*, written just a few months later, Gamow described how Bethe worked out a solution during the train journey from Washington back to Cornell University in New York, where he worked. This was a typical Gamow exaggeration; Bethe actually completed the calculations back at Cornell. At almost the same time, slightly earlier in 1938, another German physicist, Carl von Weizsäcker, working in Berlin, came up with the same idea. Bethe, however, went on to do more work on nuclear fusion inside stars, eventually (in 1967) receiving the Nobel Prize 'for his contributions to the theory of nuclear reactions, especially his discoveries concerning the energy production in stars'. Von Weizsäcker followed a different path, most contentiously, as a member of Werner

Heisenberg's team working on nuclear weapons research in the Second World War.

The idea that both of them came up with involves nuclei of carbon, nitrogen and oxygen, as well as protons. This in itself indicates that the idea was a child of its time, the 1930s, when it was still thought that about two-thirds of the mass of the Sun consisted of heavier elements than hydrogen and helium.* The model is variously known as the carbon cycle, the CN cycle or (my preference) the CNO cycle. Our understanding of it has been slightly (but only slightly) improved since 1938; what follows is a summary of the modern version of the cycle.

There are a few additional pieces of knowledge you need in order to understand the CNO cycle. First, the chemical properties of an element are determined by the number of protons in the nucleus of an atom of the element, which is equal to the number of electrons that form the cloud surrounding the nucleus: the visible face the atom presents to other atoms. But different versions of the same element, called isotopes, have different numbers of neutrons in the nucleus. At the simplest level, hydrogen can exist in a form with a single nuclear proton, or in a form with both a proton and a neutron (known as deuterium, or heavy hydrogen). Carbon can exist in several varieties, each with six nuclear protons and six electrons. Some (the most common kind) have six neutrons per nucleus ('carbon-12', since the nucleus contains 12 nucleons), some have seven neutrons per nucleus ('carbon-13'), and there are others. Secondly, a neutron can undergo a transformation in which it is converted into a proton plus an electron, which escapes at high speed. But the electron is not in any sense 'inside' the neutron; the mass-energy of the neutron is converted into proton + electron in a

---

* As I have mentioned, with a characteristic disdain for the conventions of other disciplines, astronomers refer to all elements heavier than hydrogen and helium as 'metals'.

process known as the weak interaction. An analogy might be made with a caterpillar turning into a butterfly; the butterfly is in no sense 'inside' the caterpillar before the metamorphosis. Similarly, a proton can be converted into a neutron by the reverse process, in effect absorbing an electron, or by spitting out a positively-charged particle called a positron, which is a kind of mirror-image of an electron (an example of antimatter). Positrons were only discovered in 1932, which is one reason why an understanding of the nuclear fusion processes that operate inside stars was not developed until later in the 1930s. Finally (for now), there is another kind of particle to take into account, called the neutrino. It plays a crucial role in the weak interaction, which converts protons into neutrons and neutrons into protons. But neutrinos have very little mass and interact only very feebly with other forms of matter, so although their existence was predicted on theoretical grounds as early as 1930, they were not detected until 1956. That detection was, of course, a triumphant confirmation of the accuracy of the theory.

So, now we can understand Bethe's insight from 1938. It begins with a nucleus of carbon-12 at the heart of a star. This absorbs a proton, through the tunnel effect, and becomes a nucleus of nitrogen-13. But this nucleus is unstable and spits out both a positron and a neutrino, converting itself into another isotope of carbon, carbon-13. One of the protons in the nucleus has metamorphosed into a neutron. The carbon-13 nucleus then absorbs another proton, becoming a nucleus of nitrogen-14, which absorbs yet another proton and becomes a nucleus of oxygen-15. Like nitrogen-13, oxygen-15 is unstable and decays by emitting an electron and a neutrino, becoming a nucleus of nitrogen-15. Once again, one of the protons in the nucleus has metamorphosed into a neutron. Now comes the finale. The nitrogen-15 nucleus absorbs yet another proton, but immediately ejects an alpha particle – two protons and two neutrons, a nucleus of helium-4. What is left behind is a nucleus of carbon-12, ready to act as a catalyst for the cycle to repeat. This means that, whatever astronomers thought about the composition of stars in the1930s, you only need a small amount of 'metals' to do the job, because no carbon is used up in the process. And, of course, very many carbon-12 nuclei are involved in cycles like this simultaneously. The overall effect, each time, is that four protons have been converted into two protons and two neutrons – four hydrogen nuclei converted to one helium nucleus – plus a couple of electrons and neutrinos, with energy released along the way.*

There is, though, an intriguing side-effect of this process. I said that no carbon is used up, but this is strictly only true once the cycle has reached equilibrium. Some of the reactions in the cycle proceed more quickly than others, and the slower interactions result in a kind of dam, where nuclei of one particular variety build up, like water building up behind a dam, until a balance is struck between the number of nuclei being made in the previous step and the number of nuclei 'spilling over the dam' and being converted into other nuclei in the next step. Because of the different rates of the reactions, the equilibrium occurs when the relative proportions of the elements involved are 5.5 per cent carbon-12, 0.9 per cent carbon-13, 93.6 per cent nitrogen-14, and 0.004 per cent oxygen-15. So even if a star starts out with no nitrogen at all, it will quickly build up to become the dominant participant, in terms of mass, in the CNO cycle. This is because the rate at which nitrogen-14 is converted into nitrogen-15 is much slower than the rate at which oxygen-15 is converted into nitrogen-14. So the CNO cycle is a very important source of nitrogen in the Universe – including, as we shall see, the nitrogen in the air that we breathe. The nitrogen in the air was once part of the CNO cycle inside stars that have long since died.

---

* A small proportion of the nuclei get involved in other interactions, which need not be elaborated here.

There was only one snag with Bethe's brilliant insight. Although the calculations showed that these interactions could just operate at the temperatures known to exist inside the Sun, they would be too rare (requiring fast-moving particles from way out in the high energy tail) to contribute very much energy. The CNO cycle does, however, operate efficiently enough to be the principal energy production process operating inside stars which are rather more massive than the Sun, and rather hotter in their cores. This deficiency in the CNO cycle (as far as the Sun was concerned) wasn't immediately clear in 1938, and for more than a decade after, but before the year was out Bethe and his colleague Charles Critchfield had worked out the details of an alternative energy source, which would turn out to be the main source of the Sun's energy. They jumped off from Atkinson's discovery that the fusion of two protons is the most likely nuclear fusion process to occur inside the Sun itself. Logically enough, this process is known as the proton–proton (or p–p) chain.

The p–p process starts when two fast-moving protons collide almost head on, and through the tunnel effect get close enough for the strong force to overwhelm electrical repulsion and tug them into a tight embrace. This results in one of the protons being converted into a neutron, with the resulting deuterium nucleus spitting out a positron and a neutrino. In the next link of the chain another proton tunnels into the deuterium nucleus, producing a nucleus of helium-3 (two protons plus one neutron). Finally, two nuclei of helium-3 collide and fuse, almost immediately ejecting two protons and leaving behind a nucleus of helium-4 (two protons plus two neutrons).* Just as in the CNO cycle, the overall effect is that in every chain like this four protons are converted into one helium-4 nucleus, with energy being liberated along the way. But, crucially, the p–p chain can

---

* A small proportion of the helium-3 nuclei actually get involved in more complicated interactions, but, as with the side loops of the CNO cycle, there is no need to go into the details here.

operate efficiently enough at the temperatures known to exist inside the Sun to account for the Sun's output of energy. Both processes, which convert hydrogen into helium, are known to astronomers as examples of hydrogen 'burning', although this is not burning in the everyday sense of something combining chemically with oxygen (such as hydrogen burning in oxygen to make water). Nuclear 'burning' releases vastly more energy than chemical burning. The CNO cycle is a major contributor to stellar energy production for stars with central temperatures greater than about 20 million K, which occur inside stars with at least one-and-a-half times the mass of our Sun; the p–p chain is relatively efficient at temperatures as low as 15 million K. But the watchword is 'relatively'. As mentioned, inside the Sun just one proton in every hundred million is moving fast enough to get the chain started, and not every collision even of these fast-moving protons results in fusion. As the realisation grew that the Sun really is largely composed of hydrogen, this gave astronomers a vastly greater potential timescale to contemplate – and the geologists an opportunity to say 'we told you so'.

Rocks of ages

Moving on to the modern understanding of the composition of the Sun, the rate at which the p–p chain can liberate energy tells us how long a star like the Sun, initially made mostly of hydrogen, can keep shining in a more or less steady fashion, before so much hydrogen has been converted into helium that it must change its structure and appearance. This immediately tells us that the Sun as we know it has a potential lifespan of, in round numbers, 10 billion years. The timescale problem has indeed disappeared. But how far through that 10-billion-year lifespan is the Sun today? This is where the geologists and radiochemists come in to the story.

The two key discoveries about radioactivity made by Ernest Rutherford and Frederick Soddy were that the process involves the transformation of one element into another, and that for each radio-active element there is a characteristic timescale of decay, called the half-life. For each radioactive element, the decay process also produces a distinctive mixture of other elements, known as the decay products. Some of these decay products may themselves be radioactive, so that further decays occur. Once enough radioactive processes had been studied in the laboratory, it became possible to take a naturally occurring sample of material, such as a piece of rock, and, by measuring the proportions of decay products in it, work out which radioactive elements were there long ago, even if they had all decayed. It is now even possible, given the right cir-cumstances, to work out how long ago those original radioactive elements were present – that is, how old the rock is.

Some radioactive elements have very short half-lives and are never found naturally on Earth today; others, including uranium and thorium, have such long half-lives that they are still around in measurable quantities today, even though they have been decaying ever since the Earth formed, out of the debris (we now know) of previous generations of stars in which those elements were forged. If a piece of rock contains some original uranium, say, and a mixture of this and its decay products, such as radium, then the amount of each element present can reveal the age of the rock. The key is the proportion of each element present, such as lead, to the remaining amount of radioactive material, such as uranium. The beauty of the technique is that it is independent of the actual amounts of mater-ial present, as long as there is enough to measure; all that matters is the relative amount of each element present. The resulting ages depend on factors such as the way the rock sample was formed – for example, by volcanic action – but obviously the Earth itself must be older than the oldest rock ages measured in this way.

The technique was pioneered by Rutherford himself, and by Bertram Boltwood, an American chemist, in the first decade of the

20th century. As early as 1904, Soddy, then working with William Ramsay in London, measured the rate at which the decay of uranium produces helium. Rutherford, then working in Canada, realised that this was an example of alpha decay, with the alpha particles produced by the decay (equivalent to helium nuclei) each picking up a couple of electrons from their surroundings to become atoms of helium. So he took a piece of uranium ore and measured both the amount of leftover uranium it contained and the amount of helium trapped in the rock. Making the rather optimistic assumption that none of the helium had escaped from the rock since it formed, he was able to work out an age for this particular piece of rock. It came out as 40 million years. But Rutherford knew perfectly well that this was only the minimum possible age for the rock (and by implication, for the Earth), since some helium must have escaped over the history of our planet. Nevertheless, this was an important pointer to what might be achieved by the radioactive dating technique.

Boltwood was inspired to take up the quest after hearing Rutherford give a talk on the subject at Yale University, also in 1904. Boltwood knew that the decay of uranium also produces radium, not just helium, and in 1905 he discovered that the decay of radium ultimately produces lead. By measuring the proportions of elements along this chain of decay from uranium via radium to lead he was able to estimate the ages of various samples of rock. His first estimates, produced in 1905, ranged from 92 million years to 570 million years. Unfortunately, these ages were all wrong, because they were based on flawed measurements and an inaccurate estimate of the half-life of radium. But by 1907 these teething troubles had been ironed out, and reliable estimates for the ages of various samples were coming in, ranging from 400 million years to a staggering 2 *billion* years, more than ten times Kelvin's timescale – which still held sway among astronomers at that time, in spite of

George Darwin's speculations. But, as is so often the case with this kind of breakthrough, the numbers were treated with suspicion by most geologists, and, as Rutherford and Boltwood each moved on to other things, radioactive dating was not taken seriously until the painstaking work of the British physicist Arthur Holmes established its reliability beyond reasonable doubt.

Holmes studied at the Royal College of Science in London, starting his degree in 1907. In his final year, as an undergraduate project, he dated a piece of Devonian rock from Norway, coming up with an age of 370 million years. After graduating, Holmes took a job as a geologist in Mozambique for six months in order to pay off his student debts, before returning to the Royal College (which by now had metamorphosed into Imperial College) where he obtained a doctorate in 1917. He then worked as a geologist in Burma until 1924, when he returned to academic life, first as professor of geology at Durham University and later at the University of Edinburgh. As the author of an influential textbook, *Principles of Physical Geology*, and an early proponent of the idea of continental drift, he was one of the most influential geologists of the 20th century.

During his time at Imperial College, Holmes dated many samples of rock using the uranium-radium-lead technique, and found that the oldest had an age of about 1.6 billion years. He also, in 1913, became the first person to apply the radioactive dating technique to fossils, establishing absolute dates for the fossil record for the first time. As he gradually built up an impressive weight of evidence, using difficult and tedious – but in his hands, precise – techniques, the geological community slowly came to accept the evidence for the great age of the Earth. In 1921, a debate held at the annual meeting of the British Association for the Advancement of Science established a consensus among the geologists, botanists, zoologists and physicists present, who agreed that the Earth must be a few billion years old, and that radioactive dating provides the best guide to its age. Five years later, a report from the National Research Council of the US National Academy of Sciences endorsed the technique and the age, officially adopting the 'radiometric timescale'. Since then, refinements of the technique have pushed back the age of the oldest known sample of terrestrial material (at present, small crystals of zircon from Western Australia) to about 4.4 billion years. This is in striking agreement with the ages of the oldest material found in meteorites (rocks from space that have fallen to Earth), around 4.5 billion years. Since meteorites are thought to be part of the debris left over from the formation of the Sun and Solar System, all the evidence clearly points to the Sun and its family of planets, including the Earth, having formed 4.5 billion years ago. This means that the Sun is roughly halfway through its life as a hydrogen-burning star. Where, though, did the primordial radioactive elements that give us the radiometric timescale come from? As I have already hinted, they were made inside stars. But just how they were made did not become clear until well into the 1950s.

From the Bomb to the stars

The first understanding of the CNO cycle and the p–p chain came at the end of the 1930s, just before the Second World War. Although research on 'pure' science took a back seat to the application of science for military purposes during the war years, immediately after the war astronomers' understanding of the nuclear processes going on inside stars leapt forward, not least because of the spinoff from research into the behaviour of interacting atomic nuclei carried out in connection with the atomic bomb project.* The key player in this development was Fred Hoyle, then a young researcher from the University of Cambridge, who worked on radar for the British

---

* The 'atomic' bomb is really a nuclear bomb, of course, but it's too late now to change the popular name.

Admiralty during the war years. It's an insight into Hoyle's character that although, after graduating in 1936 at the age of 21, he completed all the scientific requirements for a PhD, he never bothered to complete the bureaucratic formalities to obtain the degree.* Although he became a lecturer in mathematics in Cambridge in 1945, he remained plain 'Mr' Hoyle until his appointment as a professor in 1958. It wouldn't happen today!

In the autumn of 1944, Hoyle visited the United States and Canada as a member of a party on official Admiralty business connected with the radar work. He managed to fit in a side trip to the Mount Wilson Observatory in California, to catch up on the latest astronomical research, and also met some of the scientists involved in the bomb project. Although they were forbidden to tell him exactly what their work involved, with his scientific background and astute brain Hoyle was able to get a pretty good idea by comparing what they were allowed to talk about with which subjects were off limits. Back in England, with a quiet spell away from work during the Christmas and New Year period, Hoyle mulled over what he had picked up on the trip. From astronomer Walter Baade in California he had received the latest thoughts on the most violent stellar explosions known at the time, supernovae. From his meetings with the nuclear physicists, interpreting what they had *not* said, he had picked up the idea that a plutonium bomb could be triggered into exploding only if the plutonium were violently compressed, in a so-called implosion. The simplest way to think of this is that a critical amount of plutonium is surrounded by a shell of explosive material, which sends a shock wave moving inwards that squeezes the plutonium until the nuclei 'split' in a runaway fission process, releasing energy.

---

* According to Hoyle, by remaining technically a student he could avoid paying income tax on a grant he was receiving at the time.

Hoyle wondered whether a supernova might be the same kind of process writ large, with the implosion of a massive star collapsing under its own weight when hydrogen burning stopped, triggering a wave of nuclear interactions which then blew the star apart. He was able to calculate how this would release nuclear energy, and to get a rough idea of the proportions of the different elements that would be produced in such an explosion at different temperatures. The next step was to compare the results of the calculations with the real world.

In March 1945, Hoyle found an excuse to visit Cambridge and look up the data for the abundances of different elements on Earth, guessing that the composition of our planet must be representative of the Universe at large, apart from the absence of the hydrogen and helium found in stars. He found that when the numbers are plotted as a graph, there is an overall decline in abundance as the elements get heavier, but that for iron and related 'ferrous' metals there is a pronounced peak. This exactly matched his calculations, provided that the temperatures inside such exploding stars reached billions, not millions, of degrees. Wartime conditions delayed publication of this discovery, but it appeared in 1946 under the title 'The Synthesis of the Elements from Hydrogen'. By then, Hoyle had established to his own satisfaction that stars are mostly composed of hydrogen – one of the first astronomers to fully appreciate this. There was still a long way to go to flesh out the full story, but this was the beginning of understanding where the elements from which we are made were themselves made – understanding that we are literally stardust. When Hoyle gave a lecture on the subject to the Royal Astronomical Society in London, one of the members of the audience was Margaret Burbidge (then Margaret Peachey), of whom more shortly. She recalled:

> Listening to Fred's presentation, I sat in the RAS auditorium in wonder, experiencing that marvellous feeling of the lifting of a veil of ignorance as a bright light illuminates a great discovery.[16]

Although it took more than ten years to complete the task, the 1946 date is important historically, because it pre-dates Hoyle's involvement with what became known as the steady-state model, mentioned earlier. Although the difficulty of making heavy elements in the Big Bang was indeed a reason to take the steady-state idea seriously in the late 1940s, Hoyle's work on stellar nucleosynthesis pre-dated, and was independent of, the steady-state idea.

The last should be first

Hoyle's proposal did not immediately meet with widespread acclaim – indeed, it was largely unnoticed. After the war, with his attention focussed on establishing himself as a lecturer at the University of Cambridge, he did not immediately follow up the idea. When he did – with a bright idea that turned out to be the key to understanding how stellar nucleosynthesis works in ordinary stars before they explode, or die in a less spectacular fashion – he was delayed by a bizarre accident. One of his duties as a lecturer was to supervise research students working for their PhDs (even though he technically did not have one himself!).* Part of this role involves suggesting a good problem for the student to work on, and in 1949 Hoyle steered one of his students towards the idea of developing Bethe's work on the conversion of hydrogen into helium to find a way to convert helium into carbon inside ordinary stars, at temperatures much lower than those Hoyle had used in his investigation of the physics of supernovae.

This was an intriguing problem, because it was already well known then that elements whose nuclei have numbers of nucleons that are multiples of four are relatively common. This includes carbon-12 and oxygen-16. It is as if the nuclei have been built up by sticking

---

* One of Hoyle's research students later supervised me, so I am one of Hoyle's 'academic grandsons'.

1. (*above*) Robert Wilson (l) and Arno Penzias (r) in 1978 in front of the Crawford Hill Antenna, which revealed the existence of the cosmic background radiation

2. (*below*) George Gamow

3. (*right*) Georges Lemaître

4. Henrietta Swan and Annie Jump Canon
outside Harvard Observatory

5. Ejnar Hertzsprung

6. Cecilia Payne-Gaposchkin

7. Henry Norris Russell

8. Ernest Rutherford, 1926

9. Arthur Eddington

10. Fred Hoyle

11. Hermann Bondi

12. Solvay Conference, June 1958

Seated l–r: W.H. McCrea, J.H. Oort, G. Lemaître, C.J. Gorter, W. Pauli, Sir W.L. Bragg, J.R. Oppenheimer, C. Moller, H. Shapley, O. Heckmann; *Standing, l–r:* O.B. Klein, W.W. Morgan, **F. Hoyle** *(back)*, B.V. Kukaskin, V.A. Ambarzumian *(front)*, H.C. van de Hulst *(back)*, M. Fierz, A.R. Sandage *(back)*, W. Baade, E. Schatzman *(front)*, J.A. Wheeler *(back)*, **H. Bondi, T. Gold,** H. Zanstra *(back)*, L. Rosenfeld, P. Ledoux *(back)*, A.C.B. Lovell, J. Geheniau

13. (*left*) Alexander Friedmann, near Moscow

14. (*right*) Willem de Sitter

15. Hendrik Lorentz with Albert Einstein, circa 1920

16. Vesto Slipher

17. (*below, top*) Milton Humason, 1923

18. (*below, bottom*) Mount Wilson Observatory under construction, 1904

19. 100-inch Hooker telescope,
Mount Wilson Observatory

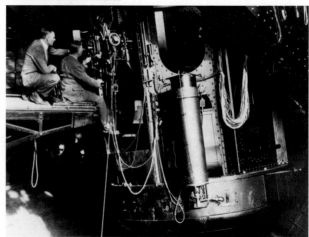

20. Edwin Hubble (l)
and James Jeans (r) at
the 100-inch telescope
at Mount Wilson
Observatory

21. 200-inch Hale
telescope, Palomar
Observatory

22. Hubble in the 200-inch observer cage, 1950

23. Allan Sandage

together nuclei of helium-4. The first hypothetical step is to stick two helium-4 nuclei together to make beryllium-8, then add another helium-4 to make carbon-12, and so on. This 'helium burning' would release energy, like nuclear hydrogen burning, but less dramatically. Faced with the task of calculating all the reaction rates up this chain of nuclear interactions as far as oxygen, the student soon became disheartened and gave up. But he did not cancel his registration as a PhD student, and, until he was forced officially to abandon any pretence that he might return to the problem, etiquette prevented Hoyle from working on the puzzle himself, or giving it to someone else. By 1952, astronomers at other universities (notably Edwin Salpeter, at Cornell) were starting to sniff around the problem independently.

The idea that the elements had been manufactured inside stars began to gain currency as astronomers began to measure the ages of stars (as described in the next chapter), and it was discovered that old stars contain fewer heavy elements than young stars – they have lower 'metallicity', in astronomical jargon. The natural explanation is that young stars are laced with 'metals' manufactured inside old stars, which have somehow dispersed these elements through interstellar space. It began to look as if someone might beat Hoyle to it. But then he got a lucky break. The recalcitrant student had finally departed, and Hoyle was invited to spend the first part of 1953 visiting Caltech and Princeton. He planned to lecture on the stellar nucleosynthesis problem, and set to work on calculating the rates of the reactions involved. He quickly found that carbon – and therefore all the elements heavier than carbon – could only be manufactured inside stars in very special circumstances.

The problem is that beryllium-8 is unstable and quickly splits apart to release two helium-4 nuclei (alpha particles). During the very brief existence of a beryllium-8 nucleus formed from a pair of colliding helium-4 nuclei, it might be hit by another alpha particle; it is so unstable that this ought to smash the beryllium-8 nucleus apart, not stick to it to make carbon-12. Yet, hypothetically, if beryllium-8 were stable, the calculations say that the production of carbon-12 would proceed so rapidly that the star would explode! Seemingly caught between the devil and the deep blue sea, Hoyle found a way to strike a balance between the options of no carbon and too much carbon, provided that the carbon-12 nucleus has a property called a resonance, with a very specific energy, 7.65 million electron volts (MeV), associated with it.

An atomic nucleus can exist in what is known as its ground state, with minimum energy, or it can absorb certain precise amounts of energy (quantised, like everything else in the subatomic world) which raise it to different energy levels. Once 'excited' in this way, it will, sooner or later, get rid of the extra energy, probably in the form of a gamma ray, and fall back to its ground state. The energy levels are like steps on a staircase, with nuclei jumping from one step to another (first up, then down) if suitably excited (like a child playing on a real staircase). Hoyle's insight was that an excited nucleus of carbon-12 could form from the collision of a helium-4 nucleus with a beryllium-8 nucleus, if (and only if) there was a step on the carbon-12 energy staircase corresponding to the combined energy of a beryllium-8 nucleus and the incoming helium-4 nucleus. It would be like tossing a ball from the bottom of a staircase with just the right speed for it to come to rest on a high step without bouncing, before gently rolling back down the stairs. This was the 7.65 MeV resonance that Hoyle predicted. If the resonance existed, the beryllium-helium interaction could manufacture carbon nuclei in the excited state, which could then radiate the excess energy away and settle into the ground state. But if the resonance did not exist, there would be no carbon, and since we are a carbon-based life form, we would not be here.

Hoyle convinced himself that, although there was no experimental evidence for such an excited state of carbon-12, it must exist. As he was at Caltech, he took the calculations to the American experimental physicist William A. ('Willie') Fowler and asked if he could do the experiment to test for the existence of this energy level. Fowler thought the idea was ridiculous, but Hoyle pestered him until he agreed to do the experiment anyway, 'in order', he later told me, 'to get Fred to shut up and go away'.* Hoyle has said it took ten days for Fowler and his team (which notably included Ward Whaling) to prove him right, contrary to their expectations; more reliable accounts say it took three months.[17] Either way, he *was* proved right.

This was a sensational discovery, the importance of which cannot be overestimated. From the fact that carbon exists – indeed, from the fact that *we* exist – Hoyle had predicted what one of its key properties must be, and opened the way to a complete understanding of how the elements are manufactured inside stars. Hoyle took a major first step even before he left Caltech in the spring of 1953, with the first draft of a paper that would be published in 1954 under the title 'I. The Synthesis of Elements from Carbon to Nickel'. But there never was a paper 'II'. Instead, he developed the full story, in collaboration with Fowler and the husband-and-wife team of Geoffrey and Margaret Burbidge, in an epic tour de force published in 1957, which also drew on parallel work by the Canadian Alastair Cameron. The authors of that paper were listed alphabetically, as Burbidge, Burbidge, Fowler and Hoyle, and it is known to this day as B²FH ('B squared F H'). Fowler received the Nobel Prize in 1983, largely

---

* I knew all of the protagonists in this story, but, as a very junior researcher in Cambridge in the late 1960s and early 1970s, I never probed them about it as much as I now wish I had. Fowler was one of the two examiners for my PhD; the other was Bill McCrea, one of the first people to realise the importance of hydrogen in the composition of the Sun. Fortunately, the Burbidges, Fowler and Hoyle were later interviewed by Ken Croswell, and their own insights can be found in Chapter Nine of his book, *The Alchemy of the Heavens*.

for this work. As even Fowler himself acknowledged privately, it should have gone to Hoyle, who seems to have missed out because he had been openly critical about previous decisions by the Nobel Committee, which exacted its petty revenge.[18] Never has there been a clearer example of the last being first in science. But that is water under the bridge. What matters is the insight the team provided into how stars work.

Stardust

This is not the place to go into all the details,[19] but the temptation to sketch the outline is irresistible. It starts with stars slightly more massive than our Sun – the Sun itself is not massive enough to allow for the production of elements heavier than carbon. A star which, like the Sun, is maintaining its output of energy by hydrogen burning, conforms to the mass–luminosity relation discussed in Chapter One, and is said to be on the main sequence. When the star has exhausted the supply of hydrogen in its core, it can no longer hold itself up against gravity, and will start to shrink. But this shrinking releases gravitational energy, which makes the centre of the star hotter, and when it reaches a temperature of around 100 million K, this triggers the conversion of helium nuclei to carbon, stabilising the star once again for as long as the supply of helium lasts. When all the available helium is used up, the star will shrink again. For the Sun, and stars with less mass than the Sun, this is the end of the story, and the star ends its life as a cooling ball of carbon nuclei (and some oxygen nuclei, since some oxygen is produced along with carbon during helium burning), surrounded by a shell of helium nuclei and a thin atmosphere of hydrogen. It has become a white dwarf, a star roughly the size of the Earth with a bit less mass than the Sun has today.

But for more massive stars, after helium burning has finished, further shrinking and corresponding temperature increases can trigger successive phases of nuclear burning. The process becomes more complicated as heavier nuclei are involved in the interactions, and nuclei which are not composed of whole numbers of alpha particles can be formed by absorbing neutrons from their surroundings or emitting positrons, which is why it took B^2FH a couple of years to work out all the details, and why elements such as nitrogen-14 exist. But in broad terms carbon burning (which occurs at a temperature of about 500 million K) produces neon, sodium and magnesium; oxygen burning (at about 1,000 million K) produces silicon, sulphur and other elements. The most important of these is silicon-28, which is eventually converted into iron through a complex series of interactions. But everything stops at iron and its closely similar element nickel. Iron-56 is the most stable arrangement of protons and neutrons in a nucleus, with the least energy per nucleon.

The elements manufactured in each stage of this process are not, though, completely destroyed in the next stage of burning. Each phase of nuclear burning (after the initial hydrogen-burning phase) takes place in a shell surrounding the core, with the shells nested like onion skins (an image first developed by Hoyle). So in an old, massive star, an iron core will be surrounded by a silicon-burning shell, an oxygen-burning shell, a carbon-burning shell, a helium-burning shell and a hydrogen-burning shell, with all the by-products of the various burning processes. Astute readers will have noticed that there is something missing from this description. Actually, two somethings – very light elements and very heavy elements.

There is far more helium in the Universe than the stars could have produced, and, in light of the work by Gamow and his colleagues, the obvious explanation was that it must have been made in the Big Bang. Although Hoyle favoured the alternative steady-state model, he was quite prepared to consider other possibilities, and described his approach to tackling scientific problems as 'compartmentalisation'. He told me that he liked to follow up one line of research without it being coloured by any preconceptions or prejudices from another area. One result of this was that, although he never abandoned the steady-state idea, he provided some of the key evidence to support the Big Bang idea. First, working with Roger Tayler in the early 1960s (the work was published in 1964), he figured out, in detail, how the proportion of helium that we see in the Universe at large could have been manufactured from hydrogen under the conditions that would have existed had there been a Big Bang. Then, he turned his attention to other light elements. Lithium, beryllium and boron are all light elements that would be destroyed at the high temperatures in the hearts of stars, but that are detected in the atmospheres of stars. The B^2FH paper could not account for their presence. Further research showed how beryllium and boron could be formed in the interstellar clouds from which new stars are born, by interactions of heavier nuclei with high-energy particles, known as cosmic rays (some of which come from supernova explosions). But in 1967 Hoyle, together with Robert Wagoner and Willie Fowler, showed, among other things, that deuterium and lithium could be produced in the right proportions under Big Bang conditions. This work made a deep impression on me. As an MSc student at the University of Sussex, I visited Cambridge to hear Wagoner give a talk on the subject. Before that talk, I still felt that the Big Bang idea and the steady-state idea were equally viable explanations for the way the Universe was; after the talk, I reluctantly abandoned the steady-state model.

The heavy elements were less of a problem, even in 1957. Making heavier elements involves an input of energy from an implosion – the supernova process that got Hoyle thinking about stellar nucleosynthesis in the first place. Although some details remained to be worked out, the overall picture was already clear. The elements manufactured inside the star are scattered through space in these explosions or, for stars with less mass than that required to trigger the explosion, more gently as an old star puffs off its outer layers. The resulting mixture of material laces interstellar clouds of hydrogen and helium, from which new stars, planets and, in at least once case, people eventually form.

'Eventually' is the key word. If the material from which the Sun and Solar System formed was made in this way, it means that at least one generation of stars must have run through its life cycle and scattered the raw materials through space first. The Sun is about 4.5 billion years old, so the Universe itself must be at least a few billion years older than that. By the middle of the 1950s, the ages of stars began to indicate that cosmologists needed to revise their ideas about the age of the Universe. In fact, the limits being set by stellar timescales were much more compelling than this rough and ready estimate.

4

13.2

The ages of stars

There are two important approaches to measuring the ages of stars. One depends on an understanding of how stars change as they age – in astronomical jargon, how they 'evolve'.* The other is an extension to stars of the radiometric dating technique for terrestrial rocks pioneered by Boltwood and Holmes. Both build from ideas first developed early in the 20th century; the stellar evolution approach bore fruit first, however, so deserves pride of place here. It starts with the independent discovery by two astronomers of a way to present the relationship between the temperature of a star (or its colour, which amounts to the same thing, as we have seen) and its brightness, on a simple graph-like diagram. This turned out to be one of the most important tools in astronomy.

Hertzsprung, Russell and the diagram

The first of the two astronomers was Ejnar Hertzsprung, a Dane who trained as a chemical engineer but became interested in astronomy and worked (unpaid) at the Observatory of the University of Copenhagen from 1902 until his reputation grew to the point where he was offered a job at the Göttingen Observatory in Germany in 1909. The other was the Princeton-based American Henry Norris

---

* With their usual insouciant disregard for the conventions of other scientific disciplines, astronomers often use the word 'evolve' to describe not just the life cycle of a single star, but also a cluster of stars, galaxy or the whole Universe. I apologise on their behalf to any offended biologists.

Russell, who had the dubious distinction of dissuading Cecilia Payne from taking her discovery of the composition of the Sun at face value. Hertzsprung published papers describing his discovery of the relationship between the brightness and colour of a star in 1905 and 1907 – but in a photographic journal which astronomers did not read, so none of them noticed. Russell discovered the same thing a little later, but published his finding (in 1913) in a journal read by astronomers; he also developed the idea more fully than Hertzsprung had. Following this, Hertzsprung's contribution was soon appreciated and acknowledged, so in this case (unlike B²FH) his name comes first both alphabetically and in terms of priority, even though he was technically 'only' an amateur at the time he did this work.

In its modern form, a Hertzsprung–Russell (or H–R) diagram*
plots the colours (or temperatures, determined from the black-body rule) of stars along the horizontal, or *x*, axis of the plot, in such a way that cooler stars are to the right and hotter stars to the left. This axis may also be calibrated in terms of the spectral class of a star, a label related to its spectroscopic appearance, but all these labels are essentially equivalent for black bodies. The vertical, or *y*, axis of the plot records the brightness of the stars, with faint stars at the bottom and bright stars at the top. This is not the apparent brightness of the star as seen from Earth, but rather its absolute magnitude: defined as the brightness it would have at a distance of exactly 10 parsecs, about 32.5 light years. Obviously, we can only work out the absolute magnitude if we know both how bright a star looks in the sky and how far away it is; so the relationships revealed by H–R diagrams could not be discovered until astronomers were able to measure distances to stars. Just how they do this is explained in Chapter Five.

The extremes of the H–R diagram correspond to bright, hot, stars (top left); faint, hot stars (bottom left); faint, cool stars (bottom

* Also known as a colour-magnitude diagram.

right); and bright, cool stars (top right). The first thing that struck astronomers when they plotted many stars on a single H–R diagram was that most of the stars lie on a band running from bottom right (cool and faint) to top left (hot and bright). This is the main sequence, and the Sun is a typical main sequence star, a little more than halfway up the band. We now know that the position of a star on the main sequence depends on its mass, a discovery made by Eddington in the 1920s, and that all of them are burning hydrogen into helium in their interiors (a discovery made much later!). Since more massive stars have to burn their fuel faster to hold themselves up, they are brighter than less massive stars. So, going up the main sequence from bottom right to top left means going up in mass.

This was not, of course, obvious in the second decade of the 20th century. Understanding how stars evolve involved a lot of painstaking work and trips down several blind alleys over the next half-century, and it would be confusing to go into all the details here. What matters in terms of measuring the ages of stars is the picture that emerged, roughly by the middle of the 1960s – about the same time that Wagoner, Fowler and Hoyle were describing how the lightest elements could have been manufactured in a Big Bang.

Ashes to ashes

Ninety per cent of all the bright stars in the Galaxy lie on the main sequence of the Hertzsprung–Russell diagram. But there are some stars which are both bright and cool, which means that they must be much bigger than the Sun; since less heat crosses each square metre of the star's surface, it needs a lot of surface to be so bright. These stars are called red giants, because of their colour and size, and lie in the upper right of the H–R diagram, above the main sequence. Some stars are hot and faint, which means that they must be much smaller than the Sun; even though more heat crosses each square metre of the star's surface, it only has a small amount of surface so is not very bright. These stars are called white dwarfs, because of their colour and size, and lie in the lower left of the H–R diagram, below the main sequence.

By studying many stars at different stages in their life cycle and combining this with computer simulations (models) of what goes on inside a star, based on the known laws of physics, astrophysicists have been able to work out how the position of star on the H–R diagram changes as the star ages. They call this an evolutionary track. This is like studying many trees at different stages in their life cycle in a forest to work out the life cycle of an individual tree.

The evolutionary track of a typical star begins when a cloud of gas and dust in space – a cloud which contains the 'ashes' of previous generations of stars, in the form of traces of 'metals' – collapses under its own weight and gets hot enough inside (through the process described by Kelvin and Helmholtz) first to make the ball of gas glow and then to trigger hydrogen burning in its heart. The trigger for this collapse may involve magnetic fields, the explosion of an old star as a supernova sending a shock wave through the clouds that exist between the stars, or turbulence in the clouds. All that matters here is that clouds do occasionally collapse in this way. 'Occasionally' is the watchword; it has been estimated that, on average, only one or two new stars (certainly fewer than ten) form in the entire Milky Way Galaxy each year. When a star does form, it settles into a place on the main sequence determined solely by its weight. More massive stars are found higher up the main sequence, less massive stars are found lower down the main sequence. And how long a star stays on the main sequence also depends only on its mass, since heavier stars need to burn fuel more vigorously to hold themselves up, so they run out of fuel sooner. The masses of stars on the main sequence range from about one-tenth of the mass of the Sun (0.1 solar masses) to about 50 times the mass of the Sun (50 solar masses). Most stars are less massive than the Sun.

As I have mentioned, a star with the same mass as the Sun can stay on the main sequence, holding itself up by burning hydrogen into helium, for about 10 billion years. A star with half the mass of the Sun will be only one-fortieth as bright as the Sun, with a surface temperature of 4,000 K, but can stay on the main sequence for 200 billion years. A star with three times the mass of the Sun will be five times as bright as the Sun, with a surface temperature of 7,000 K, but will only stay on the main sequence for 3 billion years. And a star with 25 times the mass of the Sun will be 80,000 times brighter than the Sun, have a surface temperature of 35,000 K, and use up all the hydrogen fuel in its core in just 3 million years. This opens the way to a method for measuring the ages of some stars. But first, what happens to stars when hydrogen burning ends and they are forced to leave the main sequence?

The first thing that happens is that the core, now mostly made of helium, begins to shrink and gets hotter as gravitational energy is released. This ignites hydrogen burning in a spherical shell surrounding the core. All the extra heat from the core and the hydrogen-burning shell pushes the outer layers of the star away and makes it swell up, with some matter actually being lost into space in the process. Because the star is bigger, even though it is emitting more radiation than a star like the Sun, less radiation is crossing each square metre of surface than in the case of the Sun, so the surface is cooler than the surface of a main sequence star. The star moves away from the main sequence, upwards and to the right. It has become a red giant. Eventually, the core is so hot – about 100 million K – that helium burning begins. For a star like the Sun, and any star up to about two solar masses, the onset of helium burning is a sudden process, known as the helium flash, but for more massive stars the process switches on more gradually. Either way, the star settles into a state reminiscent of its time on the main sequence, but now with a helium-burning core surrounded by a hydrogen-burning shell.* Along the way, a great deal of the outer part of the star has been puffed away into space.

For a star with up to about four solar masses, this is the end of the line. When helium burning comes to an end, the star shrinks into a white dwarf, initially very hot but gradually cooling down into a dense cinder. Stars with more than about four solar masses can go through the further stages of nuclear burning described in the previous chapter, ejecting more material (stellar ash) into space either through relatively gentle processes or, in the case of stars with more than about eight solar masses, ending their lives in violent explosions, as supernovae, scattering heavy elements through the Galaxy and leaving behind a tiny, dense star, known as a neutron star. All of this contributes to our understanding of the origin of the elements in our own bodies. But for understanding the ages of stars, the one key fact is that the point in its life at which a star turns off the main sequence depends only on its mass. This means that if we had a group of stars that all formed together, at the same time, and plotted their positions on an H–R diagram, the diagram would be incomplete. It would be lacking stars at the top end, because all the stars with more than a certain mass would have already used up their core hydrogen supply and turned off the main sequence. The position of the turnoff – the mass of the last stars that were still on the main sequence – would tell us the age of the whole group of stars. Happily, there are just such groups of stars. They are called globular clusters. But working out their ages is not quite as simple as you might hope.

Globular cluster ages

As their name suggests, globular clusters are densely packed balls of stars; an individual cluster may contain hundreds of thousands,

---

* One of the people who developed the understanding of red giant 'evolution' was John Faulkner, once one of Fred Hoyle's students, who later became my own PhD supervisor.

or even millions, of stars. We know that globular clusters are old, because their stars contain very little in the way of heavy elements – they have low metallicity. The inference is that they formed not long after the Big Bang; however, they cannot represent the very first stars to form, since they do contain some metals. They must have formed from the ashes of the very first generation of stars; so the ages of globular clusters should be a bit less than the age of the Universe, defined as the time since the Big Bang. The globular clusters in our Galaxy, the Milky Way, are spread out in a spherical halo that surrounds the Milky Way itself, which is a flattened, disc-shaped congregation of stars. This is another pointer to their great age, since it is thought that they must have formed in the cloud of material which formed the Galaxy, before it had settled down into the disc we see today. Because globular clusters are far away from us (typically, many thousand parsecs, or tens of thousands of light years), compared to the size of a globular cluster itself (typically 10 parsecs, or 32.5 light years, across), all the stars in a cluster can be regarded as effectively at the same distance from us when it comes to interpreting the H–R diagram of the cluster. Within that ball of stars, there may be a thousand or more stars in a single cubic parsec of space; for comparison, there is no star as close as that to our Sun. Although there are fewer than 200 visible globular clusters, their distribution was an important clue, in the 1920s, to the nature of the Galaxy and its relationship with other galaxies, as I discuss in Part Two. Here, all that matters is their ages.

The key to understanding the ages of globular clusters is measuring their distances from us. Only by knowing their distances can we work out the actual brightness of the stars they contain – their absolute magnitudes – and locate the turnoff on the H–R diagram in terms of mass. But this requires very accurate measurements of the distances. If you think a cluster is farther away than it really is, you will estimate its brightness to be greater than it really is. And this dramatically affects your age estimates – an error of 10 per cent in distance corresponds to an error of nearly 2 billion years in age. Until recently, it was very difficult to make these distance measurements, and the resulting estimates of the ages of globular clusters were very uncertain. One technique involves analysing the light from a class of stars known as RR Lyrae variables, which are found both in globular clusters and much closer to home. These stars are each known – from studies of relatively nearby examples, of which distances can be measured in other ways – to have a brightness that is related to the way it varies – the period of its cycle of brightening and dimming. So once an RR Lyrae has been identified in a globular cluster and its periodicity has been analysed, its distance can be inferred from its apparent brightness. But the technique is not precise.

A more rough and ready technique involves taking the H–R diagram for a cluster and adjusting all the magnitudes (in effect, sliding the whole cluster closer or farther away) until the main sequence sits on top of a 'standard' main sequence worked out by studying nearby stars. The snag with this technique is that the standard H–R diagram is based on stars that contain more metals than the stars in globular clusters. This must affect the calculation, but nobody knows quite by how much. Another problem, with all these methods, is that dust in space absorbs some of the light from distant objects, which not only affects our measurements of their brightness, but can also change the colour of the light, which is a key to estimating the temperature of a star (remember that the H–R diagram is also known as a colour-magnitude diagram). This is very similar to the way dust in the atmosphere of the Earth affects sunlight at sunset and sunrise, producing red skies; the interstellar effect is also known as reddening.

With all these difficulties, it is hardly surprising that even as recently as the mid-1990s there was considerable uncertainty about the ages of globular clusters. Using the techniques described, and some subtler methods, the best that astronomers could say was that the globular clusters were somewhere between 12 and 18 billion years old, with the middle of that range, 15 billion years, regarded as the best guess. But then everything changed.

The change came thanks to an orbiting space observatory called Hipparcos, which had been launched in 1989 by the European Space Agency (ESA). Over a lifetime of four years, Hipparcos measured accurately the distances to almost 120,000 stars, using the parallax technique, which I shall discuss in Chapter Five. The team behind Hipparcos described the precision of their measurements as equivalent to using a telescope on the top of the Eiffel Tower to measure the size of a golf ball on top of the Empire State Building. The four-year observation programme generated more than a thousand gigabytes of data, gradually sent down to Earth over the lifetime of the satellite. But because of the way the data set had been produced, the astronomers back on Earth could not extract the distance of a single star on its own, but had to wait until the observation project was complete and they could get everything out at once. Even then, the data processing took nearly as long as the observation phase had, so the results from Hipparcos were not available until 1997.

Hipparcos measured directly the distances to many different kinds of stars, including RR Lyrae variables and ordinary main-sequence stars. This had important implications in many branches of astronomy and cosmology, some of which I touch on later in this book. The most important result of the Hipparcos survey, however, was that it improved the estimates of the ages of globular clusters, both by revising the 'best guess' and by narrowing down the range of possible error surrounding that guess. It turned out that globular clusters are significantly farther away than had been thought before 1997, and, therefore, that they must be brighter than had been thought. If the stars are brighter, that means they have been burning their fuel more vigorously; so, in order to explain their present appearance, they must be younger than we had thought. A younger, hotter star has processed its nuclear fuel more rapidly than a cooler, fainter star would have. The best estimate for the ages of globular clusters came down, as a result of Hipparcos, to a range from around 10 to around 13 billion years, with 12 billion years being a reasonable rule-of-thumb value. More recently, Brian Chaboyer and Lawrence Krauss, two members of the Hipparcos team, have looked at all the available techniques for determining the ages of globular clusters, and come up with what they call a 'best fit' age of 12.6 billion years for the oldest globular clusters in the Galaxy. Happily, this ties in very well with estimates of the ages of very old stars, based on completely different techniques.*

White dwarf ages

The next technique is one that the Comte de Buffon, or even Isaac Newton, would have appreciated, if they had known about the life cycle of a star. It is very closely related to the idea of calculating the age of a lump of cooling iron by measuring its present-day temperature. The cooling 'lumps', in this case, are white dwarf stars.

A white dwarf is a star at the end point of stellar evolution, when all nuclear burning in its interior has ceased. It is essentially a hot ball of carbon, with no source of internal heat and nothing to do except sit quietly and cool off for the rest of eternity. We can calculate the age of a white dwarf if we know how hot it was to begin with (which is calculated from our models of stellar evolution, and turns out to be roughly 200,000 K to 250,000 K), how quickly it

---

* In December 2013 ESA launched Gaia, a successor to Hipparcos. It has a planned mission length of five years and will be able to measure parallaxes down to 0.0001 arcseconds (10 microarcseconds), corresponding to a distance of 100,000 parsecs, about 320,000 light years. It is hoped that Gaia will measure the parallax of more than a billion stars.

cools and its temperature today. The calculation is simplified because there is a fairly limited range of possible masses to consider. Any star which has a mass more than eight times that of the Sun explodes as a supernova and leaves behind a neutron star, in which rather more mass than there is in the Sun is squeezed into ball only a few kilometres across (about the size of Mount Everest) – it cannot be a white dwarf. Any star with much less mass than the Sun will still be sitting on the main sequence today or (as we shall see) will have become a red giant. The oldest white dwarves* around today all have leftover masses between about half the mass of the Sun and roughly three-quarters the mass of the Sun. Their outer layers, including all the 'metals', have been blown away into space. The only thing that has to be measured is the brightness, or luminosity, and temperature; the fainter (cooler) the star, the older it must be.

Working out how such a star cools might sound a daunting task, but the structure of a white dwarf is very simple, and the temperature inside it is almost the same all the way from the centre to the surface.** The cooling process is also very simple, except for a few minor subtleties which occur along the way and can be easily calculated. At the start of its time as a white dwarf, for example, the star may shrink a little, releasing gravitational energy as heat; at a certain point along the way the interior of the star crystallizes, which also releases a little heat. After this solidification it cools slightly more quickly than it had before, in a way which is well understood in terms of known physics. The result is a theoretical 'cooling curve', a graph comparing the age of a white dwarf star with its surface temperature, from which the age can be read off once you know the temperature.

---

* Astronomers often use the plural 'dwarfs', but I am a traditionalist, at least in this case.

** The theory underpinning this approach was pioneered, in the early 1950s, by the British astronomer Leon Mestel, with whom I later shared an office.

There are other subtleties, which I won't go into, but one of the key results of the calculation is a prediction of the relative number of white dwarves that should be seen at each luminosity. The actual 'white dwarf luminosity distribution' observed for stars in the disc of the Milky Way closely matches the theoretical prediction, except for a paucity of very faint white dwarves. The obvious reason for this lack of very faint white dwarves is that no stars in the Milky Way are old enough to have reached this stage of their evolution. And the very clear 'edge' to the distribution tells us that the oldest white dwarves in the disc of the Milky Way have been cooling off for 9 billion years. The masses of the stars involved in this survey are around 0.8 solar masses, and stellar evolution calculations tell us that the parent stars from which they evolved spent about 300 million years going through their life cycles. This gives an age for the disc of the Milky Way of 9.3 billion years, give or take a billion years or so. But these are not the oldest white dwarf stars in our Galaxy.

As I have already hinted, there are two stellar components to the Milky Way Galaxy. The Milky Way itself, which is a flattened disc of stars, is surrounded by a spherical halo, which includes the globular clusters (more of this in Part Two of the book). What matters here is that the halo stars formed before the Milky Way did, and are older than the disc stars. So if we can identify white dwarves in globular clusters, or elsewhere in the halo that surrounds us, we should be able to identify some of the oldest stars in the Galaxy. The snag is that halo stars are, by and large, far away, and the stars we are interested in are interesting precisely because they are very faint, which makes them hard to observe. But it can be done.

Provided they can be identified and their light analysed, white dwarves in globular clusters provide an independent way to measure both the distances and the ages of the clusters. But the technique had to wait for the advent of the Hubble Space Telescope and, in particular, an ultra-sensitive camera known as WFPC2, installed during a servicing mission in 1993. Even then, white dwarves could only be studied in a few of the nearest globular clusters to us. The difficulty in making these observations is highlighted by the fact that the apparent brightness of these stars, as seen from Earth, is less than one-billionth of the apparent brightness of the faintest stars visible to the naked eye. In order to get enough light from them to analyse, the camera had to make many exposures over several days, accumulating photons from the source over all that time.

The analysis of the light painstakingly gathered in this way was made slightly easier because the atmospheres of white dwarves are either pure hydrogen or dominated by helium. There are no metals to worry about. The structure of the star's atmosphere depends on the strength of gravity at the surface of the star, and this affects its spectrum. With sufficiently accurate measurements of the spectrum, it is possible to work out both the strength of the star's gravity (and therefore its mass) and its surface temperature. The age of a globular cluster is then revealed by the ages of its oldest, faintest white dwarves.

At the beginning of the 21st century, observations of white dwarf stars in the globular cluster M4, about 5,600 light years away from us,* came up with an age of 12.1 billion years, with a possible error range of plus or minus 0.9 billion years. A couple of other globular clusters have been studied in the same way, and are found to have similar ages. All these measurements tie in beautifully with the best fit age for the oldest globular clusters found by the Hipparcos team, 12.6 billion years. This is very encouraging evidence that astrophysicists know what they are doing – and there is more to come.

A distance of 5,600 light years counts as nearby for a globular cluster. But what if there were a few old white dwarves *really* nearby,

---

* In the direction of, but far beyond, the constellation Scorpius.

as astronomical distances go? Then we could measure their ages much more easily and even more accurately. Happily, two such stars are now known. The first one, known as SDSS J1102, was cautiously identified as a nearby 'old halo white dwarf candidate' in 2008, using observations made by the Sloan Digital Sky Survey mapping project. By 2012, its status had been confirmed and another such object, WD 0346, identified. Both stars are members of the halo population, but just happen to be passing (rather rapidly) through our neighbourhood. Indeed, it is their speed which first attracted the attention of astronomers, and which confirms their status as halo stars; disc stars move in more or less circular orbits around the centre of the Galaxy, like runners going round a track, but interlopers from the halo are expected to zip through this orderly traffic at high speed and any old angle. At present, J1102 is 50 parsecs above the disc of the Galaxy, and WD 0346 is 9 parsecs out of the disc.

J1102, which lies in the direction of Ursa Major, is moving across the sky at a rate of 1.75 seconds of arc per year, while WD 0346, which lies in the direction of Taurus, is moving at 1.3 seconds of arc per year. For comparison, the angular diameter of the Moon as seen from Earth is 30 arc minutes, or 1,800 arc seconds. This means that J1102 moves across an angle in the sky equivalent to the diameter of the Moon in just over a thousand years. This is spectacularly fast, compared with the motion of other stars across the sky, and means that the star must be both nearby and fast-moving. It is so close that its distance can be measured directly, using the parallax technique which I will describe later, and this puts it just over 100 light years away (about 34 parsecs), less than 2 per cent of the distance to M4. And in order to be moving across the sky at 1.75 seconds of arc per year, its actual sideways speed must be about 260 km per second (nearly 600,000 miles per hour). With the precise distance known, all the parameters describing J1102 (in particular, the absolute magnitude) can be very tightly constrained. The parallax measurement for WD 0346, equally valuably, places it at a distance of 28 parsecs, slightly closer than J1102, moving across the sky at 150 km per second (more than 300,000 miles per hour).

It turns out that J1102 is a 0.62 solar mass white dwarf, with a surface temperature of 3,830 K. WD 0346 has a mass slightly bigger than that of J1102 – 0.77 solar masses – and a surface temperature of 3,650 K. Allowing for the time the stars spent on the main sequence, as well as their cooling times, the total ages for the two stars come out as just under 11 billion years for J1102 and 11.5 billion years for WD 0346. These ages confirm that the stars belong in the halo and are not part of the disc of the Milky Way; they also match closely the ages of the oldest globular clusters determined by Hipparcos and by white dwarf measurements. As a bonus, studying these nearby stars will make it possible to improve the understanding of such objects, and improve the measurements of ages for globular cluster white dwarves. But this still isn't the end of the story of stellar ages.

Radiometric ages and the oldest known star

There are several candidates for the title 'oldest known star in the Galaxy', because of the uncertainty inherent in all these very difficult measurements and their interpretation in light of our best theories of stellar evolution. The various estimates made in the present century all overlap with one another, in the range of around 13 billion to 14 billion years. In itself this is a profound and significant discovery that would have both amazed and delighted previous generations of astronomers. As yet, however, it is impossible to pin down which is the actual oldest known star. What follows is what we know about a couple of the candidates, and my own 'best buy' in light of present knowledge. By the time you read this, there may be other candidates, but, hopefully, this discussion will enable you to judge for yourself how much confidence should be placed in any particular claim.

The first candidate is a relatively nearby star, dubbed HD 140283, which is just in the process of turning off the main sequence on its way to becoming a red giant. This places it at exactly the stage of its evolution that depends most sensitively on its age. Because it is so close (only some 60 parsecs, or 190 light years, away according to parallax measurements made using the Hubble Space Telescope*), the light from the star is not affected by the reddening problem, making one fewer thing to worry about. It is also close enough to be seen with the aid of good binoculars, if you look in the right place in the constellation Libra. But, like the nearby white dwarf stars WD 0346 and J1102, HD 140283 is just visiting our neighbourhood; it is a high-speed traveller from the halo, crossing the sky at an astonishing 0.13 milliarcseconds per hour, fast enough that its motion can be seen in photographs taken by the HST a few hours apart. This is equivalent, at the distance the star is from us, to a speed through space of about 350 km per second (roughly 800,000 miles per hour). Indeed, HD 140283 was identified as an unusually fast-moving star as long ago as 1912, and was the first star ever shown, by spectroscopy, to have far fewer heavy elements than the Sun – to be 'metal deficient', in astronomical terms. This was first a clue to its great age, and then the means of measuring that age. From studies of its orbit, astronomers infer that the star was probably born in a small 'dwarf' galaxy that approached too close to the Milky Way for its own good, and got shredded by tidal effects as its stars were sucked by gravity into elongated orbits, repeatedly diving in close to the Milky Way and climbing far out into the halo.

'Metals' comprise about 1.6 per cent of the Sun by mass. Astronomers measure the metallicity of a star by comparing the proportion of elements like iron relative to hydrogen revealed by their

---

* The HST measures individual stellar parallaxes very accurately, but, unlike Hipparcos, cannot measure simultaneously large numbers of parallaxes.

spectra. The metallicity of the Sun is defined as 1, and the metallicity of other stars is sometimes measured in terms of units called 'dex', short for 'decimal exponent'. On this scale, 1 unit corresponds to a factor of 10; so if there is 10 times as much iron (relative to hydrogen) in a star as there is in the Sun, it has a dex of 1; 100 times as much iron corresponds to a dex of 2, and so on. If the metallicity is less than that of the Sun, the units are negative – a dex of −1 means one-tenth as much as the Sun, a dex of −2 means one-hundredth as much as the Sun, and so on. The metallicity of HD 140283 is about 1/250th of the Sun's metallicity.

As well as measuring the metallicity of a star like HD 140283 compared with the Sun, astronomers can measure the proportions of different heavy elements within the star itself. These depend on the age of the star, which determines how much of each element has been manufactured by nucleosynthesis; the relative proportions of oxygen and iron in particular provide a good indication of age. In HD 140283, the dex of oxygen is −1.5, and the dex of iron is −2.3. Using this and other evidence, in 2013 a team headed by Howard Bond, then working at Pennsylvania State University, reported a face value age for the star of 14.5 billion years. This produced newspaper headlines describing it as 'the oldest star', but that is not quite the whole story. There is considerable uncertainty in this age determination, both because of the difficulty of making the observations and the uncertainty of details of the theory underpinning the calculation. Increasing the oxygen dex slightly (by about 0.15), but still within the range of possible errors of those measurements, would, for example, reduce the implied age of the star to 13.3 billion years. The effect of a trace of reddening would also reduce the calculated age. So, at present the best estimated age for HD 140283 is 14.5 plus or minus 0.8 billion years – that is, anything in the range from 13.7 to 15.3 billion years. This displaced, in a sense, a star known as CS 22892-052 as the 'oldest known star', but it is worth mentioning

CS 22892-052 to show how dramatic changes in our understanding of such objects have been in recent decades. I have a fondness for the directness of the method of dating stellar ages used for CS 22892-052, the last one that I shall describe, and the way it harks back to the earliest direct measurements of the age of the Earth.

In 1996, when I was writing *The Birth of Time*, some difficult and supremely accurate spectroscopic studies of this star – involving the measurement of abundances of many elements, but in particular thorium and europium – had been used to measure an age of 15.2 billion years, plus or minus 3.7 billion years. By 2003, further studies of the same star, combining observations from the ground and from the HST, had refined this to a thorium/europium age of 12.8 ± 3 billion years, and an estimate based on several different elements of 14.2 ± 3 billion years. All of these estimates agree with an actual age in the range of 13 to 13.5 billion of years, near the lower end of the estimates for HD 140283, and thorium/europium dating of other stars gave similar results at the beginning of the present century. But how does it work?

If the Comte de Buffon and Isaac Newton would have had no trouble understanding the principle behind measuring white dwarf ages, Bertram Boltwood and Arthur Holmes would have been completely comfortable with the last method I shall describe for determining the ages of stars. It is simply radiometric dating applied on an astrophysical, rather than a geophysical, scale. The white dwarf technique applies to stars that started their lives with more mass than the Sun and have evolved more rapidly; the radiometric technique involves stars that started out with less mass than the Sun, have evolved more slowly and are, in spite of their great age, in the red giant phase of their lives today.

In Chapter Three, I glossed over the fact that elements come in different varieties, called isotopes, which have different masses (because they have different numbers of neutrons in their nuclei) but the same chemical properties (because they have the same number of protons, and therefore the same number of electrons). Ordinary hydrogen and deuterium (heavy hydrogen) are different isotopes of hydrogen, and helium comes in two varieties: helium-3 and helium-4, the first with two protons and one neutron in each nucleus, the second with two protons and two neutrons. This is important in radiometric dating, because for some heavy elements there are both stable isotopes and unstable isotopes. So, when we talk about the radioactive decay of an element we really mean the decay of a particular isotope of that element.

There is a rough and ready way to calculate the age of the Milky Way from radiometric dating, which gives us one profound result. The proportions of different isotopes around today can tell us about the proportions of radioactive isotopes around at the time the Solar System formed – even the ones that have long since decayed, since we can identify and analyse the isotopes produced by those decays that are still around today. So we know roughly what mix of radioactive elements was around in the clouds of interstellar space at the time the Solar System formed, and we can use those estimates to work out when that mixture of material formed. The simplest guess is that it all formed in one go, at the birth of the Milky Way. This is obviously wrong, because we know there are still supernovae going off today. But it is wrong in a very useful way. It sets the minimum possible age of the Milky Way, which turns out to be 8 billion years. There is no way that the Galaxy can be younger than this, and therefore no way that the Universe can be younger than this – a point to bear in mind as we move on to Part Two of this book.

A slightly more educated guess is that the same number of supernovae have gone off every year (or rather, every thousand years, since they only go off at a rate of one or two per century) since the Milky Way formed, lacing the clouds in space with fresh radioactive materials and other stuff. That probably gives us an overestimate of the age, since supernova activity was probably higher when the Milky Way was young. It gives an age of roughly 13 billion years, with an uncertainty of plus or minus 3 billion years, happily close to the ranges of the ages of individual old stars. Which brings me, at last, to my best buy.

The last breakthrough I want to describe was the detection of spectroscopic features produced by uranium-238 in stellar spectra. The thorium used in previous age measurements (thorium-232) has such a long half-life, 14.1 billion years, that very little of it has decayed even on the timescales we are talking about here. This half-life is, for example, more than three times the age of the Earth. With such a long half-life and very little decay even on astronomical timescales, very few of its decay products are around, making them hard to find and analyse. Astronomers knew that uranium-238, with a half-life of 'only' 4.5 billion years (roughly the same as the age of the Earth) and with well-understood, easily-identified decay products, would provide a much better cosmic clock, if only they could find traces of this isotope in the spectrum of a star, or stars. The breakthrough came at the beginning of 2001, when a team of astronomers using a telescope at the European Southern Observatory high in the mountains of Chile reported the discovery of the tell-tale lines produced by uranium-238 in the spectrum of a star known as CS 31082-001. The star has a thousand times less iron than the Sun (dex −3), and contains thorium as well as uranium, making it doubly valuable as a clock. The relative proportions of thorium and uranium in the star today give a good indicator of its age, which comes out as 12.5 billion years, give or take 3 billion years. Not quite the oldest star known, but (at the time) one of the oldest studied by the technique I regard as the most reliable. Then, in 2008, came HE 1523-0901.

HE 1523-0901 is a red giant halo star with about 80 per cent of the mass of the Sun, about 7,500 light years away in the direction of the constellation Libra, with a dex of −2.95. Anna Frebel, then at the University of Texas at Austin, and her colleagues reported that they had not only identified spectroscopic features caused by uranium and thorium in the light from the star, using the Very Large Telescope at the European Southern Observatory, but also other elements including europium, osmium and iridium. This enabled them to measure a whole slew of ratios: uranium/thorium, thorium/iridium, thorium/europium and thorium/osmium. The more such ratios there are to analyse, the more reliable the age estimate is. Putting it all together, the team came up with an age of 13.2 billion years, with an uncertainty of about plus or minus 3 billion years. This is slightly more than the 'face value' age of CS 31082-001, but there is a twist in the tail – the small difference between the uranium/thorium ratios for CS 31082-001 and HE 1523-0901 suggests that the former is actually slightly older than the latter, regardless of their absolute ages, and this is certainly allowed by the range of uncertainties in the estimates. But, says the team: 'given that the observational uncertainties exceed [the difference], the present ages of the two stars suggest that they formed at roughly the same time. This is also reflected in their almost identical metallicity.'

The bottom line is that all these ages, calculated in three different ways – globular cluster ages, white dwarf cooling ages and radiometric ages – agree with one another. This tells us two things. Astrophysics works – astronomers know what they are talking about. And the oldest stars in the Galaxy are a bit more than 13 billion years old. How does this fit in with our understanding of the Universe at large?

PART TWO

How Do We Know
the Age of the Universe?

5   **31.415**

Prehistory: Galaxies and the Universe at large

Our local neighbourhood of space is dominated by stars. But we now know that this is because we live on an island of stars, the Milky Way Galaxy, and that on larger scales the Universe is dominated, at least visually, by galaxies. Although stars are obvious to anyone who looks at the night sky, because of their great distances even relatively nearby galaxies appear only as fuzzy blobs of light in the sky, difficult to see at all without the aid of a telescope. So it is hardly surprising that the first European description of these fuzzy blobs, then known as nebulae, appeared only in 1614, shortly after the invention of the telescope, in the writings of Simon Marius. He was a German astronomer who had 'Latinised' his name (as was the scientific fashion in Europe at the time) from Simon Mayr. As well as being the first Western scientist to identify what we now know as the Andromeda Galaxy, which had previously been known to Arab scholars, Marius discovered the four main moons of Jupiter at about the same time as Galileo, but did not publish news of his discovery at once.* It was, however, another hundred years before Edmond Halley, of comet fame, brought the study of the nebulae into the astronomical mainstream with a paper published in the *Philosophical Transactions* of the Royal Society in 1716. His explanation of the phenomenon was, though, incorrect:

---

*   Nevertheless, the names by which those moons are now known (Io, Europa, Ganymede and Callisto), are the ones given to them by Marius.

> Not less wonderful are certain luminous Spots or Patches, which discover themselves only by the Telescope, and appear to the naked Eye like small Fixt Stars; but in reality are nothing else but the Light coming from an extraordinary great Space in the Ether; through which a lucid medium is diffused, that shines with its own proper Lustre.

What Halley did not realise is that many of these nebulae (galaxies) are made of stars, and shine for that reason; this was to be a stumbling block in the development of an understanding of nebulae for the next two centuries, not least because it turns out that there are two kinds of nebula. The ones I am interested in here are indeed other galaxies, more or less like the Milky Way. But there are also clouds of gas and dust between the stars in the Milky Way that shine, in many cases, because they contain hot stars. The famous nebula in the constellation Orion is an example. Indeed, the Orion Nebula is the first nebula in a list compiled by Halley; the second nebula on his list is the Andromeda Galaxy. Today, the term 'nebula' is reserved for these clouds of material, and 'galaxy' for the objects once known as nebulae that lie beyond the Milky Way. For clarity, I shall sometimes use the word 'galaxy' where Halley and his successors would have said nebula.

One thing Halley did get right was his realisation that because these objects, unlike the planets, are not seen to move among the stars, they must be at very great distances. And because they are seen as extended objects, not as tiny points of light like stars, that meant they must be very big. This inspired some prescient, but premature, speculations about the size and scope of the Universe.

The power of pure reason

It began with the work of the 18th-century thinker Thomas Wright, of Durham. In 1750, he published a book with the splendid title, *An original theory or new Hypothesis of the Universe, founded upon the laws of nature, and solving by mathematical principles the general phenomena of the visible creation, and particularly the Via Lactea.*[20] The 'Via Lactea' is, of course, the Milky Way. The book was a mixture of good and bad, with philosophy, theology and science jumbled up in something of a confusion, but it contained one powerful idea. Wright suggested that the appearance of the Milky Way as a band across the sky could be explained if all the stars formed a disc, like a mill wheel, with the individual stars 'all moving the same Way, and not much deviating from the same plane, as the Planets in their heliocentric Motion'. In this picture, the stars move around the centre of the Milky Way in the same way that the planets move around the Sun; Wright went further by suggesting that other stars would have their own families of planets orbiting around them in the same way. And if there could be other solar systems (or as he more accurately said, 'sidereal systems'), why not other Milky Ways? Using the word 'Creation' where we would say 'galaxy', he went on: 'As the visible Creation is supposed to be full of sidereal Systems and planetary Worlds, so on, in like manner, the endless Immensity is an unlimited Plenum of Creations.' In other words, an infinite Universe populated with innumerable galaxies like the Milky Way. The nebulae, he specifically says, 'may be external Creations'. Which led him to speculate on the insignificance of humankind in the cosmos:

> In this great Celestial Creation, the Catastrophy of a World, such as ours, or even the total Dissolution of a System of Worlds, may possibly be no more to the great Author of Nature, than the most common Accident in Life with us, and in all Probability such final and general DoomsDays may be as frequent there, as even Birth-Days or Mortality with us upon this Earth.

This is a somewhat surprising speculation from someone who did believe in the existence of a Creator.

These ideas were picked out of the confusion of Wright's book by the philosopher Immanuel Kant, and were the inspiration for his attempt to go one step further, explaining the behaviour of the observed Universe in terms of Newton's laws, without invoking the hand of God. By 1755, Kant had written a book in which he presented a much more scientific view of the nebulae as 'island universes', explaining that disc-shaped systems of stars could appear circular if seen face-on but elliptical if viewed at an angle. He subscribed to the idea of a boundless, infinite Universe, and suggested that the Universe we see around us had evolved from an earlier state. Unfortunately, these ideas did not receive widespread attention at the time, because Kant's publisher went bust and the book was never properly distributed. So it was left for Pierre-Simon Laplace to spread (and get credit for) the idea of what became known as the 'nebular hypothesis' in his *Exposition du Système du Monde*, in 1796, and more fully in the fifth volume of his *Traité de Mècanique*, in 1799. He said that the nebulae must contain billions of stars, like the ones in the Milky Way, and that the Milky Way itself would look just like one of these nebulae if seen from a great distance. In other words, we do not live in a special place in the Universe. This book is also famous for his description of what are now known as black holes, and for Laplace's response to Napoleon's enquiry as to why the book did not mention God: '*Sire, je n'avais pas besoin de cette hypothèse-là.*'* But this was about as far as theory and pure reason could take the debate at the end of the 18th century. What was needed now were more and better observations, which came during the 19th century, but in a confusing way.

One step forward, two steps back

The first significant step had already been taken by the time Laplace published his ideas. In the mid-1780s, the pioneering astronomer and

---

* 'Sir, I had no need of that hypothesis.'

telescope maker William Herschel reported a series of observations of nebulae made with his new reflecting telescope. With an aperture of just over 18 inches and a focal length of 20 feet, this was more powerful than anything in use before. Using this instrument, he was not only able to see many more nebulae (nearly 500 by 1784), but was also able to resolve some of the objects previously classified as nebulae into clusters of stars. Because of their appearance as round balls packed with stars, some of these became known as 'globular' clusters; others have a looser structure and are known as 'open' clusters. All those clusters had to be part of the Milky Way. But Herschel was, initially, in no doubt that the unresolved nebulae lay beyond the Milky Way. In 1785, he said that some of these nebulae 'may well outvie our milky-way in grandeur', and speculated that stars had originally been spread evenly across the Universe, but had joined together to make nebulae (galaxies) because of gravitational attraction. In 1786, he wrote:

> To the inhabitants of the nebulae of the present catalogue, our sidereal system must appear either as a small nebulous patch; an extended streak of milky light; a large resolvable nebula; a very compressed cluster of minute stars hardly discernible; or as an immense collection of large scattered stars of various sizes. And either of these appearances will take place with them according as their own situation is more or less remote from ours.

He said that the Milky Way must be separated from the nebulae by large tracts of empty space, and he tried to calculate the size of the Milky Way.

This effort was bedevilled by a problem which would cause confusion right into the 20th century. Unknown to Herschel, there is dust between the stars in the disc of the Milky Way, and this obscures the light from very distant stars. The effect is rather like being in a diffuse fog. If you are standing in a field in a fog, you can only see for a limited distance in any direction, and it looks as if you are in the centre of a small circular field. When the fog lifts and you can see farther, it may turn out that you are actually in one corner of a large, square field. In the same way, because of interstellar dust, we seem to be in the centre of the disc of the Milky Way; but, as we shall see, modern observations (including those at infrared wavelengths which penetrate the dust) show that we are a long way out towards the edge of the disc of our Galaxy. Nevertheless, Herschel deserves credit for making the effort, even if his estimates of the size of the Milky Way were far too small – a disc with a diameter, in modern units, of about 2,200 parsecs and a thickness of about 520 parsecs.

But then Herschel took a backward step. He had previously recognised that there are different kinds of nebula – some far beyond the Milky Way, some within the Milky Way and some, the so-called planetary nebulae, which 'leave me almost in doubt where to class them'. Planetary nebulae get their name because in a small telescope they appear as circular patches of light, like planets, not as points of light, like stars. We now know that they are clouds of material puffed off by stars in the late stages of their lives, post main sequence. It was Herschel who first saw the evidence for this. In 1791, observing an object which became known as NGC 1514 with a new 40-foot telescope, he saw what seemed to be 'a star which is involved [that is, embedded] in a shining fluid, of a nature totally unknown to us'. As he found more objects like this, Herschel backed away from the idea of nebulae as other galaxies. He became fascinated by the idea that planetary nebulae might be the sites of star birth (exactly the opposite of the truth!), and in 1811 wrote that although he had previously 'surmised nebulae to be no other than clusters of stars disguised by their very great distance', longer experience 'will not allow a general admission of such a principle'. The next great observational advance muddied the waters further still.

In 1845, William Parsons, the 3rd Earl of Rosse, completed the construction of a huge telescope on his estate at Birr Castle in Ireland. It was so big that it became known as the 'Leviathan of Parsonstown' and was not superseded in size until the construction of the 100-inch Hooker Telescope in 1917. The Leviathan had a mirror 72 inches (1.8 metres) across, five inches (13 cm) thick and weighing three tons. The rest of the telescope was in proportion, and the 54-foot (16.5 m) tube, weighing roughly twelve tons, could be moved through a large angle up and down (in altitude) and to a more limited extent sideways (in azimuth). The driving force for the construction of the telescope was Rosse's fascination with nebulae, which he was wealthy enough to go to extreme lengths to satisfy. He set out to study as many of them as he could. It was Rosse who discovered that some nebulae have a spiral shape; by 1850 he had found fourteen spiral nebulae, leading him to write, in a paper published by the Royal Society, 'as observations have accumulated the subject has become, to my mind at least, more mysterious and more inapproachable'.

There was, indeed, now a confusion of nebulae. Some were spiral, some elliptical, some were planetary nebulae, some (such as the Orion Nebula) just seemed to be glowing clouds within the Milky Way. Rosse gave up trying to explain what was going on. But a breakthrough was just around the corner.

Nebular spectroscopy

Rosse died in 1867, the year before Janssen and Lockyer discovered the spectroscopic features caused by the presence of helium in the Sun. Three years earlier, in 1864, the first breakthrough in the spectroscopic study of nebulae had been achieved. William Huggins, another private astronomer, who had built an observatory in south London, had been intrigued by news of Kirchhoff's discoveries, and with the aid of a neighbour, William Miller, set out to analyse the spectra of stars and nebulae. They found similarities between the spectra of stars and the spectrum of the Sun; but Huggins then made the discovery, published in 1864, that the spectra of planetary nebulae did not contain the lines characteristic of stellar spectra but were almost featureless, like the spectrum of a cloud of gas. He did, though, find star-like features in the spectra of other nebulae, including the spiral (as it was now known to be, thanks to Rosse) in Andromeda (M31).

By 1866, the year before Rosse died, Huggins had enough data to make a groundbreaking contribution to the annual meeting of the British Association for the Advancement of Science, held that year in Nottingham. He reported that many nebulae, including the planetary nebulae, are gaseous, even though individual planetary nebulae may harbour a single star in their heart. But all the objects originally classed as nebulae that could now be resolved into stars using modern telescopes (notably the globular clusters) had, hardly surprisingly, an overall spectrum like that of an individual star. Significantly, though, many of the nebulae that could not be resolved into individual stars, including Rosse's spirals, had spectra similar to those of globular clusters. All the evidence pointed to these nebulae being agglomerations of stars, too far away for the individual stars to be resolved, although Huggins stopped short of stating this definitively.

While astronomers still hesitated to accept the idea of island universes, galaxies beyond the Milky Way, technology continued to nudge them in that direction. In the second half of the 19th century, photography was beginning to supplant the human eye in the study of the heavens. Instead of having to observe an object and draw what could be seen, astronomers could photograph it, producing a more accurate image that could be studied at leisure. There was another advantage. Once your eye has adapted to the dark, you can stare at an object for hours, but never see anything you couldn't see in the first few minutes. A photographic plate, however, continues to absorb light and build up an image for a very long time. This makes it possible to see more detail than can be seen with the human eye, and even to photograph things that cannot be seen at all with the human eye. With a spectroscopic camera attached to an astronomical telescope, accurate spectra of faint objects can be preserved for posterity, and studied using a microscope to pick out tiny details of the lines in the spectrum.

One of the pioneers of this technique was Julius Scheiner, working at the Potsdam Observatory. He obtained a spectrum of the Andromeda Nebula by exposing a plate in such a camera for seven and a half hours. It amply confirmed Huggins' findings. In 1899, Scheiner reported that since 'the previous suspicion that the spiral nebulae are star clusters is now raised to a certainty, the thought suggests itself of comparing these systems with our stellar system, with special reference to its great similarity to the Andromeda nebula'. In other words, the Milky Way and the Andromeda Nebula are both spiral galaxies. The scene was set for the 20th century. The next key step would have to await the construction of the 100-inch Hooker Telescope on Mount Wilson in California, but even in the first two decades of the new century, photography and spectroscopy stimulated a revival of interest in nebulae, particularly the spirals. It set astronomers off on a long and winding road.

First steps

As Confucius said, a journey of a thousand miles begins with a single step. The first step along the road from the Earth to the Universe at large was taken in 1761, when astronomers used observations of a rare transit of Venus (when Venus passes across the face of the Sun as seen from Earth) and geometric techniques to work out the distance to the Sun. This involved making accurately timed observations of the transit – in particular, the moments when Venus seemed to be just touching the edge of the Sun's disc – from widely separated places on Earth. With the distance from the Earth to the Sun known (modern measurements give this as 149.6 million kilometres), the width of the Earth's orbit (just under 300 million km) can be used as a baseline to measure the distances to the nearest stars. This is because the nearby stars seem to shift slightly against the background of more distant ('fixed') stars as the Earth moves around the Sun. This parallax effect is exactly the same as the way your finger seems to move, if you hold it out at arm's length and shut your eyes in turn. But the shifts measured by astronomers are much smaller than that. A good benchmark is the angular diameter of the Moon, which is 30 minutes of arc, or 1,800 arc seconds. Even for nearby stars, the parallax effect is tiny compared with this. The distance for which a star would show a displacement of one second of arc in photographs taken six months apart is called a parsec, from parallax second of arc, and this is roughly 3.26 light years. The nearest star is 1.32 parsecs (4.29 light years) away, so all these measurements involve observations of displacements of less than one second of arc – in round numbers, less than one two-thousandth of the apparent size of the Moon. They were impossible before the advent of astrophotography.

There are other, less accurate, techniques for estimating the distances to open clusters of stars by studying the way they move across the sky, and the techniques based on an understanding of the Hertzsprung–Russell diagram described earlier. But the key step on the road to the Universe at large was taken at Harvard in 1912, by Henrietta Swan Leavitt, an experienced assistant to the astronomer Edward Pickering. She had graduated from the Society for the Collegiate Education of Women (which went on to become Radcliffe College) in 1892, just before her 24th birthday, and joined Pickering's team at Harvard College Observatory (initially as a volunteer) the following year. Her job was to interpret photographic plates to determine the magnitude (brightness) of stars; she became expert at interpreting the behaviour of stars which vary in brightness and measuring how much they varied. In 1896, she left for a two-year trip to Europe, but on her return Pickering offered her a paid job, and she became a full-time professional astronomer (at 30 cents an hour), a member of a team of female 'computers', as they were known.

The variable stars studied by Leavitt were initially thought to be binary systems, which varied in brightness because one star periodically passed in front of the other. But it became clear that they were actually individual stars, which really did vary in brightness over time, sometimes taking many months to complete the cycle of brightening and dimming. Although her work was often interrupted by illness, in 1904 Leavitt was the right person in the right place when she was working on a box of photographic plates that had arrived at Harvard from the observatory's southern observing station in Arequipa, Peru. The plates were photographs of a pair of nebulae only visible from the southern hemisphere, known as the Magellanic Clouds, after the first European to describe them, Ferdinand Magellan. She quickly found dozens of variables in one of these nebulae, the Small Magellanic Cloud, and even more when more plates arrived from Peru later that year. The number for both clouds soon ran into the hundreds. In 1908, she published a paper summarising her work so far, with the explanatory title '1777 Variables in the Magellanic Clouds'. The key discovery for which she is famous was mentioned right at the end of the 21-page paper: 'It is worthy of notice that the brighter variables have the longer periods.'

As Leavitt's biographer George Johnson has pointed out, this is the astronomical equivalent of the remark made by Francis Crick and James Watson at the end of their famous DNA paper: 'It has not escaped our notice that the specific pairing we have postulated immediately suggests a possible copying method for the genetic material.' Their discovery was the key to life; Leavitt's discovery was the key to the Universe.

The relevant point is that if the periods (the time from one peak of brightness to the next) of a certain class of variable stars are related to their brightness, you only have to measure the length of a star's period to know how bright it is. But there is a snag. The relationship has to be calibrated. You have to find at least one member of this particular family of stars whose distance has been determined in some way. Without this calibration, if you find similar stars to these variables in the Milky Way, by using the period–luminosity relation you can say that one star is, say, twice as bright as another, and must be correspondingly more distant to look as faint as it does to us; but you cannot tell the absolute distance to any of them. Once you know the distances to a handful of these stars, however, you know their absolute magnitudes, and you can use the periodicity measurements of other members of the family to work out their absolute magnitudes, which can be compared with their apparent brightness to give their distances. The dimming caused by distance drops out of the calculation for the Magellanic Clouds because these nebulae are (as we now know, and as Leavitt and her contemporaries surmised) so far away that it is effectively the same for all the stars in the clouds. The distance from one side to the other side of one of the clouds is only a small percentage of the distance from here to the clouds. The family of stars identified by Leavitt is now known as Cepheids, after the archetypal member of the family, a variable star in the constellation Cepheus (known as Delta Cephei), studied by the English astronomer John Goodricke in the 1780s.

Leavitt's work progressed very slowly, because of her own ill health and the death of her father in 1911. But by 1912 she had found 25 variables in the Small Magellanic Cloud that showed a close relationship between brightness and period that could be plotted as a simple graph. This was enough to put the relationship to work measuring distances across the Milky Way – if only the distance to at least one 'local' Cepheid could be measured directly. Unfortunately, there was no Cepheid close enough for its distance to be measured by parallax with the telescopes available at the time* – not even the closest one, which just happens to be the northern Pole Star. So, the crucial first step in calibrating the Cepheid distance scale was taken (by Ejnar Hertzsprung) using a more rough and ready technique called statistical parallax. This is a neat trick and is surprisingly accurate if applied to enough stars. It depends on observing a large number of stars, such as an open cluster, close enough for their movement across the line of sight to be measured by watching their position change from year to year, in terms of the angle covered. The stars will all be seen moving in roughly the same direction, but some faster and some slower. Their actual velocities towards or away from us can be measured directly, by the familiar Doppler effect, and this gives a measure of the amount of random velocities associated with the stars, moving relative to one another. It is a reasonable assumption that the actual random velocities across the line of sight are the same, on average, as the average random velocities towards or away from us; so, subtracting the randomness inferred by the Doppler measurements from the velocities across the line of sight leaves the actual velocities across the line of sight, which can be compared with the angle through which the stars are seen to move each year to give the distance.

Hertzsprung applied this technique in 1913 to measure the distances to a few Cepheids, calibrate Leavitt's distance scale and work out the distance to the Small Magellanic Cloud. The answer he came up with was 30,000 light years (nearly 10,000 parsecs) – but the number that appeared in his paper was 3,000 light years because

---

* Things are different now, thanks to satellites such as Hipparcos.

of a misprint. The 30,000 light year estimate was a mind-blowing distance to astronomers at the time. Although, for various reasons, it was not much more than a tenth of the actual distance, it provided the basis for a reassessment of the size of the Milky Way and our place in the Universe.

The long and winding road

At the beginning of the 20th century, astronomical understanding of the Milky Way had advanced little since Herschel's day – if anything, it had gone backwards. So the Dutch astronomer Jacobus Kapteyn was starting basically from scratch when, in 1906, he devised a plan to study the structure of the Milky Way by counting the numbers of stars with different apparent magnitudes, spectral types, radial (Doppler) velocities and sideways movement (proper motion) in different parts of the sky. The project used data from more than 40 different observatories and took the best part of two decades to complete. But it suffered from a major flaw. Although the presence of material between the stars was known by then, Kapteyn took little account of the effect of the resulting dimming (interstellar extinction) – indeed, the effect was not properly understood until the 1930s. So when Kapteyn published his results in 1920 they presented essentially the same 'foggy' picture of our surroundings that Herschel had described, but in more detail. The established picture of the Milky Way was still as a disc-shaped system of stars, with the Sun close to its centre; it was widely accepted that even if the Milky Way was not the entire Universe, any 'external' nebulae must be small satellites relatively close to us. But this picture was starting to change even by the time Kapteyn published his conclusions. The first dramatic step was the displacement of the Sun from the centre of the Milky Way.

The man who took this step was Harlow Shapley, then working at the Mount Wilson Observatory in California, who had been quick to build on Leavitt's discovery of the period–luminosity relation for Cepheids. Although he went up some blind alleys, which need not be discussed here, in 1918 Shapley reported that he had measured the distances to some nearby globular clusters, using the Cepheid period–luminosity relation, and used these distances to determine the brightness (absolute magnitudes) of the brightest stars in globular clusters. Since these turned out to be much the same (not surprising, since there is a limit to how big a star can be without exploding), he could calculate distances to other globular clusters by measuring the brightness of their brightest stars; then, less accurately, he could estimate the distances to even more distant clusters by assuming that they all have the same diameter, so their apparent size in the sky reveals their distance. These observations were not plagued by interstellar extinction, because the globular clusters lie above and below the plane of our Galaxy, which is where the dust is thickest. By measuring the distribution of the globular clusters in space, Shapley found that they are arranged in a sphere centred on a point in the direction of the constellation Sagittarius. This point, he reasoned (correctly) must be the true centre of the Milky Way, with the Sun and its family of planets far out in the galactic suburbs.

Shapley also used the distances he had calculated to work out the distance to the centre of the Milky Way, but here he came a cropper. We now know that the stars he used in the first step of the calculation were not Cepheids, but a similar family of variables known as RR Lyrae stars. These are intrinsically dimmer than Cepheids, so they were not as far away as Shapley thought. The result was that his calculation gave an enormous size for the Galaxy. He thought that the centre of the Milky Way was about 20,000 parsecs (roughly 65,000 light years) away, with the diameter of the whole disc being about 90,000 parsecs (300,000 light years). This was a hundred times the size of the Milky Way inferred by earlier studies. The idea of such a huge Galaxy lent weight to the idea that other nebulae are merely satellites of the Milky Way, and Shapley reinforced this argument by measuring the brightness of what he thought were novae in external nebulae.

Novae are stars that explode at the end of their lives, briefly shining much brighter than any main sequence star. But there is a limit to how bright a nova can be, well known from studies of novae in the Milky Way. If the spiral nebulae were galaxies like our Milky Way, with roughly the same huge diameter that Shapley had calculated, then in order to explain their appearance as tiny patches of light in the sky they would have be at distances of hundreds of millions of light years, much too far away for even a nova to be visible from Earth if it exploded in one of them. And yet, Shapley pointed out, novae had been observed in spiral nebulae. If those novae were the same brightness as novae within the Milky Way, that meant the spirals must be just beyond the outer fringes of his huge Galaxy. The resulting image was of a great Milky Way Galaxy, the biggest thing in the Universe, ploughing through space with an attendant retinue of small nebulae, and possibly in the process of gobbling them up. But were the novae seen in spiral nebulae he had studied really the same brightness as novae seen in the Milky Way? Alas for Shapley, by bad luck it turned out later that the exploding stars he had used in his investigation were not mere novae, but the even brighter stellar explosions, unrecognised at the time, now known as supernovae.

One astronomer who disagreed with Shapley's interpretation of the evidence was his compatriot Heber Curtis, who had a very different picture of the Universe. Their opposing views were presented at a meeting of the American National Academy of Sciences (NAS) on 26 April 1920, which became known in astronomy as 'The Great Debate'. It resolved nothing, but it set the scene for the next step out into the Universe.

An unresolved debate

During the second decade of the 20th century, working at the Lick Observatory in California, Curtis carried out a major study of spiral nebulae. From the numbers he could see in different patches of the sky, he calculated that there must be a million of these visible to the instrument he was using (a telescope with a 36-inch diameter mirror, known as the Crossley Reflector). This was a staggering number for astronomers to contemplate at the time, but far short of the number of galaxies now known to exist. From studies of the dark 'lanes' that are distinctive features of the spiral structure of these nebulae, Curtis concluded that relatively star-free regions of the Milky Way must be similar dark 'lanes' in the Milky Way, and that it is indeed just an ordinary spiral galaxy. This was borne out by his measurements of the distances to spiral nebulae, based on his studies of novae. As it happens, the novae he studied really were similar to novae in the Milky Way, not the even brighter supernovae that had confused Shapley's measurements. So, Curtis came up with distances to external nebulae that were pretty much in the same ball park as modern distance measurements – tens of millions of light years for relatively nearby galaxies. He became a leading proponent (arguably *the* leading proponent) of the 'island universe' idea and wrote, in a paper published in 1917:

> If we assume equality of absolute magnitude for galactic and spiral novae, then the latter, being apparently 10 magnitudes fainter, are of the order of 100 times as far away as the former. That is, the spirals containing the novae are far outside our stellar system; and these particular spirals are undoubtedly, judging from their comparatively great angular diameters, the nearer spirals.

So far, so good. But, like Shapley, Curtis also made a crucial mistake. He simply could not, or would not, accept Shapley's measurements of the distances to globular clusters. He agreed that they must be spread out through a spherical volume of space centred on the Milky Way, but he thought that our Galaxy was only about 30,000 light years across, with the Sun about 10,000 light years from its centre.

The disagreement between the world view promoted by Shapley and the world view promoted by Curtis provided the impetus for the discussion on 'The Scale of the Universe' organised by the NAS in 1920. Although it became known as The Great Debate, there was very little discussion involved, and no debating. Each of the champions presented his own view of the Universe, and each published separate papers, leaving it more or less up to the audience and readers to make up their own minds. And, in spite of the title of the meeting, Shapley, for one, was less interested in the scale of the Universe than in the scale of the Milky Way. Indeed, before going to Washington DC for the meeting, Shapley wrote to a colleague that he did not intend to say much about spiral nebulae because he did not have a strong argument to support his ideas. At the meeting, Shapley's principal claim was that: 'Recent studies of [globular] clusters and related subjects seem to me to leave no alternative to the belief that the galactic system is at least ten times greater in diameter – at least a thousand times greater in volume – than recently supposed.'

By contrast, Curtis's main objective at the meeting was to promote the idea that spiral nebulae are galaxies like the Milky Way, whatever their size. But he did say that 'the island universe theory had an indirect bearing on the general subject of galactic dimensions' because:

If the spirals are island universes it would seem reasonable and most probable to assign to them dimensions of the same order as our galaxy. If, however, their dimensions are as great as 300,000 light years, the island universes must be placed at such enormous distances that it would be necessary to assign what seem impossibly great absolute magnitudes to the novae which have appeared in these objects.

Those 'impossibly great' magnitudes later turned out to be not so impossible, when supernovae were identified, but Curtis can hardly be blamed for not knowing that in 1920. He also emphasised that the optical spectrum of a spiral nebula is the same as the overall spectrum of the Milky Way.

In one respect, Curtis seems to have been a little more open-minded than Shapley. He acknowledged that 'it is, of course, entirely possible to hold both to the island universe theory and to the belief in the greater dimensions for our galaxy by making the not improbable assumption that our own island universe, by chance, happens to be several fold larger than the average'. This was an idea that held sway (in a diluted form) for a surprisingly long time, partly, perhaps, through an unconscious wish for there to be something special about our place in the Universe. It was only in 1998 that a team at the University of Sussex (of which I was a member), using data from the Hubble Space Telescope, established once and for all that the Milky Way is an average spiral galaxy, at least in terms of size.[21]

In their measurements of the size of our Galaxy, Shapley's estimate was too big, and Curtis's estimate was too small. But Curtis made a much more serious mistake by placing the Sun relatively close to the centre of the Milky Way. In terms of the nature of the spiral nebulae, Curtis was right and Shapley was wrong. But there was another puzzle which made astronomers hesitant about accepting his view of the Universe.

The confusion was caused, in all innocence, by the Dutch astronomer Adriaan van Maanen, working at the Mount Wilson Observatory in California, alongside Shapley. Van Maanen

– unfortunately, as it turned out – was a friend of Shapley, and the claims that he made in the second decade of the 20th century had a great influence on Shapley's thinking. Van Maanen had been studying the spiral nebulae (in particular, one known as M101) using photographs spanning the interval from 1899 to 1915. He identified distinctive features in the nebula (brighter blobs of light) and compared the photographs from different years, using an instrument in which two images rapidly and repeatedly replace each other in the field of view, so that differences jump out to the human eye (a 'blink comparator'). Van Maanen convinced himself that in some cases these bright features had shifted by a tiny amount over the years, implying that the nebulae were rotating. The implied rotation was slow, about one turn every couple of hundred thousand years (a rotational speed of 0.02 seconds of arc per year for M101). If the nebulae were the same sort of size as the Milky Way and at the distances implied by the island universe idea (M101 is roughly half a degree across as seen from Earth, about the same angular size as the Moon), this meant that the outer regions of these nebulae must be moving faster than the speed of light! Van Maanen – and Shapley – saw this as a *reductio ad absurdum*. The nebulae could not rotate faster than light; so, they must be much smaller so that the problem did not arise, and therefore much closer – relatively nearby objects.

When other astronomers tried to reproduce van Maanen's results, they did not succeed. But van Maanen insisted he was right, and Shapley believed him. Nobody knows for sure where van Maanen went wrong, although one suggestion is that since his observations relied on measurements of the outer regions of nebulae, at the edge of the field of view of his instrument, something may have been wrong with its optics. Or it may just have been wishful thinking. Either way, sufficient doubt remained to ensure that the island universes idea remained controversial at the beginning of the 1920s – but not for much longer.

A universe destroyed

The man who killed off the idea that the Milky Way is the biggest thing in the Universe, and that the spiral nebulae are merely satellites of the Milky Way, was Edwin Hubble, a figure who looms so large in the story – partly thanks to his skill as a self-publicist – that he deserves more than a cursory introduction.

Hubble was born in 1889 and attended both high school and university in Chicago. He was a decent athlete, although not as good as he liked to pretend, but a genuinely top student. He studied not only science and mathematics, but also French and the Classics, leading him to win a coveted Rhodes Scholarship to study law in Oxford for two years following his graduation from the University of Chicago in 1910. There, he fell in love with a kind of P.G. Wodehouse version of Englishness and turned himself into a faux 'English gentleman', affecting a 'British' accent and expressions such as 'Bah Jove!', which would irritate many of his colleagues in years to come. Hubble never practised law, but after working as a high school teacher and sorting out family affairs following the early death of his father in 1913, he joined the Yerkes Observatory, near Chicago, as a research student in astronomy, gaining his PhD in 1917. His project there involved photographing as many faint nebulae as possible, using the Yerkes 40-inch refracting telescope that was one of the best astronomical instruments in the world at the time. Even before he finished this work he had been offered a job at Mount Wilson, in anticipation of the completion of the 100-inch reflector. But the United States entered the First World War that year, and Hubble asked for the job to be put on hold while he joined the infantry and went off to Europe.

Hubble had an undistinguished military career. The official records show that he arrived in France shortly before the end of hostilities and never experienced combat. That was hardly his fault, but he always gave the impression in later life that he had received a wound which affected the mobility of his right elbow. The affliction was genuine, whatever its cause. With the fighting over, Major Hubble, as he now was (and how he would always like to be addressed even in civilian life) managed to linger in his beloved England long enough to irritate his employers at Mount Wilson, where the new telescope was up and running. He arrived there in September 1919, just before his 30th birthday. There, he briefly worked alongside Shapley, before Shapley moved on to Harvard in 1921. The relationship between the two astronomers was not cordial. Shapley was a more down-to-Earth character. Hubble's affectations got right up his nose, and in return Hubble looked down his nose at Shapley.

The first significant work by Hubble was a development of his PhD research, which led to a classification of nebulae (galaxies) according to their appearance. The key contribution was his realisation that there are two kinds of nebulae, the now-familiar spirals and a separate family known as ellipticals, which have no spiral structure but come in a variety of appearances from spherical (like huge globular clusters) to cigar-shaped. It is now thought that ellipticals form from the mergers of spirals. This project was not completed until 1923, but in the course of it Hubble became an expert at coaxing the best out of the new 100-inch telescope. Then he turned his attention to the problem of measuring the distances to the nebulae.

Armed with two of the best telescopes in the world, the 60-inch and 100-inch reflectors, Hubble was ideally placed to test the island universe idea, which he found persuasive but was too cautious to go out to bat for without solid evidence. He embarked on a project to search for novae in the nebulae, and along the way by the summer of 1923 had found several variable stars in an irregular nebula known as NGC 6822. Further studies revealed eleven Cepheids in the nebula, which indicated a distance to NGC 6822 of about 700,000 light years, outside the boundary of even Shapley's supergalaxy. By the time this was established, Hubble had taken a huge leap out into the Universe.

Probably encouraged by his discovery of Cepheids in NGC 6822, Hubble intensified his search of the spirals. On 4 October 1923, although conditions were poor for observing, Hubble managed to obtain a photograph, using the 100-inch, of the Andromeda Nebula, M 31. This showed a bright speck of light within the cloudy shape of the nebula itself; 'Nova suspected', he wrote in his log book. The next night, the astronomical 'seeing' was better, and the bright speck was still visible in the photograph he took; 'Confirms nova', he wrote. A more detailed study of the photographic plate then showed not one but three suspected novae. This sent Hubble searching through the archive of photographs of the nebula to confirm that these really were 'new' stars, not something that had been seen before. Two of them did turn out to be novae. But the other one had indeed been photographed, although apparently not always understood before. It was a star whose brightness fluctuated – a variable. But what kind of variable? To find out, it had to be monitored at every opportunity.

In February 1924 this monitoring produced the evidence Hubble needed. Over three nights, he saw the star double in brightness and was able to combine these data with the data from the archive and his own earlier observations to determine the period of the variation. The star was a Cepheid with a period of 31.415 days. On 19 February, eager to crow, he wrote to Shapley with the news: 'You will be interested to hear that I have found a Cepheid variable in the Andromeda nebula', he said and, rubbing salt in to the wound, pointed out that, using the same relationship that Shapley had used to work out distances to globular clusters, this meant that the Andromeda Nebula must be a million light years away, possibly more if interstellar dimming had to be taken into account. It was indeed an 'island universe' similar to the Milky Way – and the

Milky Way, by implication, was just another galaxy, not the whole Universe. Cecilia Payne-Gaposchkin happened to be in Shapley's office soon after he received the news and recollects him saying: 'Here is the letter that destroyed my universe'. And later: 'I believed in van Maanen's results […] he was my friend.'

For Hubble, the next step was obvious. He would have to measure distances to as many galaxies as possible, and for that he would need an assistant. This project was to lead to an even more startling discovery than the insignificance of the Milky Way in the Universe at large – but a discovery which had roots that pre-dated the award of Hubble's PhD, let alone of his discovery of the distance to M 31.

6 575

The discovery of the expanding Universe

The fact that the Universe is expanding is one of the most profound discoveries in science, and leads directly to the realisation
that there was a beginning to the Universe as we know it. The first
steps towards that realisation were taken by Vesto Melvin Slipher
('VM' to his colleagues), working at the Lowell Observatory (in
Flagstaff, Arizona) in the second decade of the 20th century.

Surprising speeds

Slipher, who had been born in 1875, arrived in Flagstaff fresh from
completing his degree at Indiana University in 1901, and was set the
task of breaking in a new spectrograph by the director of the observatory, Percival Lowell. Lowell, who came from a wealthy Bostonian
family, had founded the observatory in 1894, originally inspired by
his belief that the 'canals' of Mars were a sign of intelligent civiliza-
tion at work on the Red Planet.* The new instrument was intended,
in the first instance, to measure the rotation of the planet Venus,
which also intrigued him. Planetary studies occupied Slipher for
the next few years, during which he became an expert at handling
the spectrograph. In 1906, following a suggestion by Lowell (who,
like many of his contemporaries, thought that the spiral nebulae
might be places in the Milky Way where new planetary systems, like

---

*   The observatory is known today for its association with the Discovery Channel
Telescope, actually located at an outstation 40 miles southeast of Flagstaff, the
Happy Jack site.

the Solar System, were forming), Slipher had made an unsuccessful attempt to measure spectra of spiral nebulae. But in 1909,* after hearing that other astronomers were turning their attention to the problem, he decided to make another attempt.

The equipment Slipher had to carry out his investigation was pretty modest – a 24-inch-diameter refracting telescope and the somewhat cranky (but now familiar) spectrograph. Although stellar spectroscopy was an established technique by then, obtaining spectra of faint nebulae was a much harder task, and nobody had yet succeeded in getting satisfactory results, even with bigger telescopes than the one available to Slipher. But after many months of patiently experimenting with different possibilities, in the 'spare' time not occupied by his work for Lowell, he found a way to tweak the combined telescope-spectrograph setup to obtain spectra from nebulae such as the one in Andromeda. By January 1913, helped by the acquisition of a new camera lens for the spectrograph, Slipher had four photographic plates on which spectral lines visible in the light from the nebula could be measured. To his surprise, he found that the lines were shifted towards the blue end of the spectrum, presumably by the Doppler effect, indicating that the Andromeda Nebula is rushing towards us at a speed of 300 km per second. This is far greater than the Doppler velocities of stars; hardly surprisingly, the announcement met with an initially sceptical response.

But Slipher persevered. By 1914 he had measured the spectra of fifteen nebulae and was able to report to the meeting of the American Astronomical Society in August that year that just three of these showed blueshifts, while eleven were redshifted. This was clearly an important discovery, and it was reported that Slipher received a standing ovation at the end of his talk. By then, other

---

* That same year he was awarded a PhD by Indiana University; he had been able to take time out from Lowell on various occasions to complete his graduate work, based on his astronomical research there.

observers had begun to confirm his findings. There was, though, a limit to what could be achieved with the relatively modest telescope at Slipher's disposal, and his most comprehensive formal paper on the subject, published in 1917, included only ten more nebulae, giving a total of 25, four with blueshifts and 21 with redshifts. The speeds indicated by the redshifts were in the range from 150 km per second to 1,100 km per second, which was suggestive evidence that the spirals, whatever they were, could not be in the gravitational grip of the Milky Way. By 1917, Slipher himself had no doubt about the implication:

> It has for a long time been suggested that the spiral nebulae are stellar systems seen at great distances. This is the so-called 'island universe' theory, which regards our stellar system and the Milky Way as a great spiral nebula which we see from within. This theory, it seems to me, gains favour in the present observations.

There was another curious feature of these observations, which is not always given the attention it deserves. The redshifts, if interpreted as velocities, implied that the galaxies were moving away from us in all directions. Or rather, not exactly away from us. When Slipher averaged out the velocities, it seemed that the whole family of spiral nebulae (or at least, the score or so he had investigated) was moving relative to the Milky Way – or, more logically, that the whole Milky Way Galaxy itself was moving through space like the other nebulae, in a certain direction relative to the spiral nebulae, with a speed of about 700 km per second. He described this as a 'drift through space' (some drift!). It is a profound discovery because it provided more evidence that the Milky Way is just an ordinary galaxy and, in particular, that we are not at the still centre of the Universe.

Slipher's observations did not, however, settle the question of the nature of spiral nebulae, and as we have seen the debate rumbled on into the 1920s. This was partly because there was still wriggle-room for Shapley and other proponents of the idea of a great Milky Way Galaxy, the biggest thing in the Universe, with an attendant retinue of small nebulae. All they had to do was imagine that the spiral nebulae were small objects being expelled from the Milky Way, shot out into the surrounding space. It didn't help that although Slipher continued obtaining spectra of nebulae and had measured 41 by 1922, almost all of them (36) showing redshifts, he did not publish all the details, and they languished in the Lowell archive in the form of internal reports, not widely read or appreciated in spite of being picked up by the astronomers Arthur Eddington and Gustav Strömberg. Everything changed, though, when Hubble started measuring distances to the nebulae studied by Slipher and then, with his colleague Milton Humason, the distances and redshifts to more remote galaxies.

Taking the credit

Hubble knew all about Slipher's work, and in 1928 he attended a scientific meeting in Leiden where he discussed the new theories about the Universe, based on Albert Einstein's general theory of relativity, with Willem de Sitter. (More about these ideas shortly.) Hubble also knew that the nebulae which looked smaller and fainter in the sky had larger redshifts than bigger, brighter nebulae, suggesting that if all the spirals are about the same size the redshift might be an indication of distance, with high-redshift galaxies being farther away from us. Indeed, the year before, in 1927, Hubble had instructed a more junior observer at Mount Wilson, Milton Humason, to measure the redshifts of two nearby galaxies ('nearby' as revealed by the Cepheid technique) to check Slipher's observations. Humason confirmed that the redshifts were relatively small, consistent with the idea that the closer galaxies have smaller redshifts.

Hubble was not particularly concerned about the reason for redshifts, but he was excited by the prospect of using redshifts to measure distances, because redshifts could be measured in galaxies far too faint (that is, he suspected, too remote) to use the Cepheid technique. Proving that there was a relationship between redshift and distance would involve measuring redshifts and Cepheid distances to as many galaxies as possible, pushing the 100-inch telescope to its limit. The work would be tedious and time-consuming, and Hubble would need help. If he measured distances, using Cepheids and any other tricks he could devise, while a colleague measured redshifts, he would be able to put the two pieces of the puzzle together to work out the relationship between redshift and distance. The obvious person to work with him on the project was Humason, not only because he was a first-class observer who knew the 100-inch well but because his background made him very much junior to Hubble. This meant Hubble could ensure (as he always did) that he got the lion's share (if not all) of the credit for the project for himself.

Humason was born in Minnesota in 1891, but moved to the West Coast of the USA with his family as a child. He first visited Mount Wilson on a camping holiday in 1905, the year after construction of the observatory there began. He loved the mountain so much that (amazingly, with his parents' approval) he dropped out of school and got a job as bellboy and handyman at the Mount Wilson Hotel. But this was only on the lower slopes of the mountain, and he soon moved on to become one of the drivers of the mule trains carrying the supplies required for the ongoing construction, and other equipment, up the mountain trail to the observatory. There was still plenty of construction going on after the 60-inch was completed, because it was rapidly followed by the 100-inch Hooker Telescope, funded by the eponymous John D. Hooker, a Los Angeles business-man. In 1911, Humason married Helen Dowd, the daughter of one of the chief engineers on the project, but carried on with his job until 1913, when she gave birth to their first child. He now had to set-tle down in a proper job to support his family, and after working for a time as a gardener, in 1916 he was he was able to buy a fruit farm (in California-speak, a 'citrus ranch') just outside Pasadena. It's clear, though, that Humason never settled into this way of life, because when the staff at the observatory began to expand with the completion of the 100-inch, he got a job, partly thanks to the contacts provided by his father-in-law, as a night janitor on the mountain.

This doesn't sound like much of a job, and it wasn't. But this was November 1917, the war in Europe was raging, and Humason was expected to do anything the astronomers needed, from making sure they knew how to keep the telescopes pointing in the right direction to making cups of coffee and developing photographic plates. The pay was a modest $80 a month, but he got a rent-free cabin and free meals while he was working. History does not recall how his wife felt about this change of career, but Humason proved so adept at the tasks he was given that he was soon granted the status of 'night assistant' and allowed to do some observing himself. Several astronomers, including Shapley, taught him the ropes, and one of them, Seth Nicholson, took the trouble to bring Humason up to speed with the mathematics he had missed by dropping out of school. Shapley later described Humason as 'one of the best observers we have ever had', and partly thanks to Shapley's recommendation Humason officially became an 'Assistant Astronomer' in 1922, although by then he had been acting as an observer without the title or pay for years.

This was shortly after one of the most extraordinary near-misses in the history of astronomy, which happened just before Shapley left Mount Wilson for Harvard. In those days, the photographic images of galaxies and other astronomical objects were captured on one side of fragile glass plates, coated with the appropriate chemicals. These plates had to be exposed for many hours, and handled in the dark and cold of the telescope dome before the images on them were 'fixed' using other chemicals. This left a permanent image on one side of the plate, with the back blank. So, astronomers could draw or write on the back of the glass to highlight objects of interest. Early in his career, still not officially an astronomer, in the winter of 1920–21 the 29-year-old Humason was asked (or told) by Shapley to examine a series of photographs of the Andromeda Nebula, mostly taken by Shapley over the past couple of years, to see if it was changing as time passed – in particular, to see if it was rotating. In these images, which were actually photographic nega- tives, so that bright objects showed up black, Humason found several spots which looked like stars. Even more intriguingly, some of these spots were visible on some plates but not on others, suggesting to him that they might be variable stars, perhaps even Cepheids. So he marked the back of one of these plates in ink to highlight a particu- larly interesting object, and showed it to Shapley. Shapley, convinced that the spiral nebulae were clouds of material within the Milky Way (or at most small, nearby objects), took a clean handkerchief from his pocket, wiped the marks away and patiently explained to Humason that it was impossible for variable stars to be present in the Andromeda Nebula. Humason was in no position to argue and kept quiet, only mentioning the incident many years later. So it was that Shapley did not gain fame in 1921 for discovering the distance to the Andromeda Galaxy and opening up the Universe. A salutary lesson that theories have to be based on observations, not observations on theoretical preconceptions.

Humason was not exactly delighted to be given the task of meas- uring redshifts for Hubble at the end of 1928. This meant obtaining photographic spectra (much harder than simply photographing galaxies) from long exposures in the freezing cold of winter on the mountain. This had to be done in winter to get longer intervals of darkness and in the cold because the telescope dome had to be unheated so that no swirling air currents distorted the 'seeing' of the instrument. Although the telescope did have an automatic guidance mechanism, to track the apparent movement of an object across the sky as the Earth rotates, this was not perfect, and an observer had to sit in the cage of the instrument all the time, looking through a smaller guide telescope and tweaking the mechanism to ensure that it was pointing in the right direction all the time. Even then, one night was not enough to obtain the kind of detailed image Humason needed. At the end of an observing run, the plate had to be taken out of the spectrographic camera (still in the dark) and packed away in a light-proof box ready to be brought out the next night (or the next clear night) when it would be put back (still in the dark) in the camera, and the telescope carefully pointed to exactly the same place as before for more hours of cold, tedious, eye-straining work. However much he disliked it, though, Humason was the best observer around and the right man for the job. He started measuring redshifts to fainter galaxies, beyond the range of Slipher's telescope.

Meanwhile, Hubble was measuring distances, first to the galaxies whose redshifts had already been measured by Slipher. He was able to measure Cepheid distances to six of these, and used this as a rough guide to show that the brightest stars in these galaxies were about as bright as each other. So, he could estimate the distances to more remote galaxies, where Cepheids could not be seen, by assuming that their brightest stars also had this average intrinsic brightness (absolute magnitude) and determining their distances from their apparent brightness in the sky. This gave him fourteen more distances. From this list of twenty galaxies, he worked out an average brightness for a galaxy, and used this as a rough and ready way to estimate the distances to four more galaxies. In 1929, out of these 24 galaxies, twenty had redshifts measured by Slipher and four 'new' redshifts obtained by Humason. This was enough for Hubble to publish his famous discovery of the redshift–distance relationship – that the distance of a galaxy from us is directly proportional to the velocity implied by its redshift. This became known as 'Hubble's

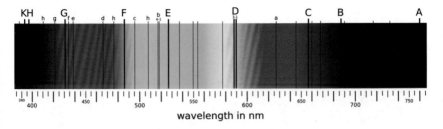

wavelength in nm

24. Fraunhofer lines

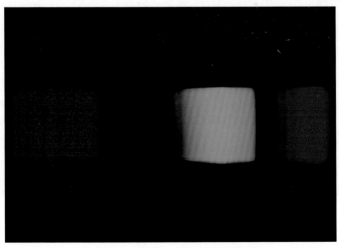

25. Flame emission spectrum of copper

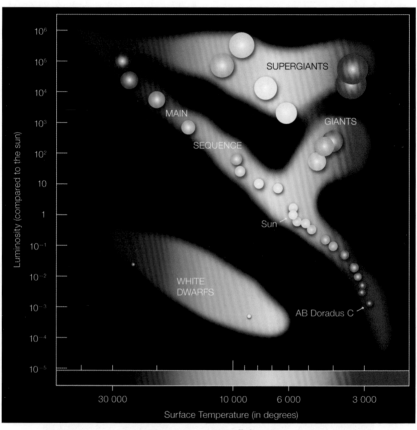

26. Hertzsprung–Russell diagram

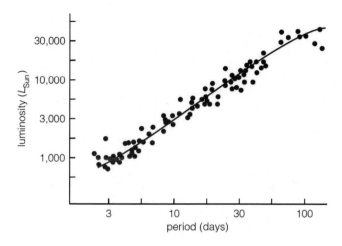

27. Leavitt's plot of brightness vs period
for Cepheids in Small Magellanic Cloud

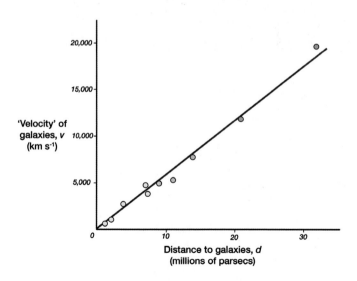

28. Hubble's law

The 'velocity' of galaxies (v) is proportional to the distance to the galaxies (d).
The two values are related by the Hubble constant (H).

29. Infrared view of the Orion Nebula

30. The Lowell Observatory at Anderson Mesa, Arizona

31. Spiral galaxy NGC 1232

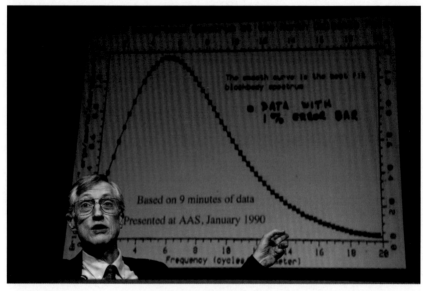

32. The echo of the Big Bang

John C. Mather shows some of the earliest data from the COBE spacecraft (the COBE blackbody curve) on 6 October 2006 at NASA Headquarters in Washington, DC.

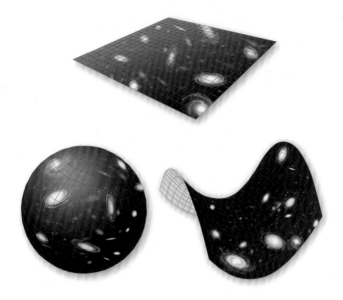

33. Flat, closed and open universe diagram

34. Cosmic microwave background seen by Planck

The map shows tiny temperature fluctuations that correspond to regions of slightly
different densities at very early times, representing the seeds of all future structure,
including the galaxies of today.

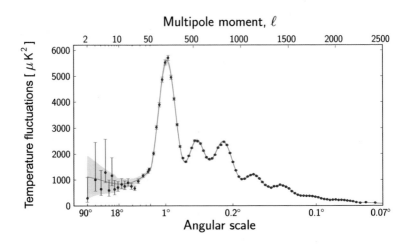

35. Planck power spectrum of the cosmic microwave background

The dots are measurements made by Planck. The wiggly line represents
the predictions of the 'Lambda–CDM' model of the Universe.

law'. He duly presented it in the *Proceedings of the National Academy of Sciences*, but with one significant tweak.

Even though Hubble's 1929 paper gave no citation to Slipher's work (itself an astonishing omission, but surely deliberate; historian Don Lago says 'there was nothing innocent about Hubble's silence', and Shapley described Hubble as 'absurdly vain and pompous'[22]), the published 'velocities' he used were tweaked by subtracting out the 700 km per second 'drift' of the Milky Way that Slipher had discovered. This left him with a velocity–distance relationship following the rule that every 500 km per second of velocity corresponded to a distance of a million parsecs (1 megaparsec, or Mpc). A redshift corresponding to a velocity of about 1,000 km per second would imply a distance of 2 Mpc, and so on. This number, 500 km per second per Mpc, became known as Hubble's constant, H, and its precise value would become a source of great debate and controversy in the years and decades ahead. But another point needs emphasising. Although the units in which redshift is measured are the units of velocity, km per second, Hubble was careful not to interpret these redshifts as being a result of the Doppler effect. He was simply interested in using them as distance indicators, and told the *Los Angeles Times* in 1929 that 'it is difficult to believe that the velocities are real'.

Once Hubble's law was published, and Hubble's constant measured, it could be used to determine the distance of any galaxy once its redshift had been measured. In a separate paper published alongside Hubble's, Humason reported the most dramatic redshift measurement to date, for a galaxy known as NGC 7619 in the direction of the constellation Pegasus. This involved an exposure over several nights, adding up to 33 hours, with a follow-up taking 45 hours. The result was a redshift measurement corresponding to a velocity of 3,779 km per second, more than twice as great as the largest redshift velocity measured by Slipher, and a distance estimate of roughly 8 Mpc, more than 25 million light years. The key significance of this breakthrough was that the Mount Wilson authorities were sufficiently impressed to fund a better spectrograph, which, with the advent of faster photographic plates, enabled Humason to push farther out into the Universe with slightly less demanding working conditions. Over the next couple of years, the survey added 40 more galaxies and reached out to 100 million light years. As the nomenclature still in use today shows, Hubble took the credit, even though Slipher pioneered the redshift work and Humason pushed it to the limit of what was possible at the time. But what did it all mean? In fact, as Hubble cannot have been unaware, even by 1928 there were already sound theoretical reasons to expect the Universe to be expanding – or at least, to expect that there is a relationship between redshift and distance.

A Russian revolution

Einstein developed his general theory of relativity at the end of 1915 and almost immediately applied it to provide a mathematical description of the entire Universe. This wasn't quite as bold a leap as it might seem, because the general theory describes the interaction between space, time and matter, and strictly speaking it applies only to a 'complete' collection of space, time and matter – that is, a universe. When it is applied to describe anything less than an entire universe,* such as the nature of the orbit of Mercury about the Sun, this is really an approximation, although the approximation can be made as accurate as you like. Einstein published his groundbreaking cosmological paper, with a title which translates as 'Cosmological Considerations Arising from the General Theory of Relativity', in 1917. He was strongly influenced by the prevailing wisdom that the Milky Way represented the entire Universe, and the observed fact

---

* By convention, 'Universe' refers to the real world in which we live; 'universe' refers to a mathematical description (a model) of a possible world, one allowed by the laws of physics but which may not be the same as the one we live in.

that the stars in the Milky Way show only relatively small random motions, with no overall outward or inward flow. He favoured the idea that the Universe is closed, in the same sense that the surface of the Earth or of any sphere is closed. A sphere has a finite surface area but no edge; a spherical universe has a finite volume but no boundary – if you head off in any direction in a straight line, you will eventually travel right round the universe and get back to where you started.

But there was one snag. Such a 'closed' universe tends to collapse, as all the matter in it pulls on all the other matter by gravity; this is true using both Newtonian theory and the general theory. So, Einstein added an extra term to his equations, called the cosmological constant and denoted by the Greek letter lambda (Λ), which acted as a pressure, or a springiness of space, pushing outwards to balance the inward tug of gravity. The result was a mathematical description of a closed, spherical universe containing matter, but stable, as required by the fact of the small velocities of the stars, as Einstein put it.

While he was developing these ideas, in 1916, Einstein discussed them with the Dutch astronomer Willem de Sitter, who soon developed his own variation on the theme. Holland was neutral in the First World War, so it was relatively easy for news to pass from Einstein in Germany to de Sitter, and then on to England – in particular, to Arthur Eddington. De Sitter published his own work in the *Monthly Notices of the Royal Astronomical Society*. This drew the attention of astronomers in the English-speaking world to Einstein's breakthrough, but it also showed that there was more to the idea than Einstein had realised. De Sitter found that the equations of the general theory could also be used to describe a stable but empty universe – spacetime without any matter in it. Hardly surprisingly, such an empty spacetime would not collapse since there was no matter in it to make it collapse. So, there was no need to invoke a cosmological constant, although you could have one if you wanted. De Sitter, however, was interested in the possibility that the presence of stars in the Universe might be such an insignificant amount of matter compared with the vastness of space that his model might be a good description of reality. He did the mathematical equivalent of sprinkling a few specks of matter into his empty universe, and came up with a surprising discovery. If these 'test particles' radiated light, then the wavelength of the light got stretched as it moved away from the particles – or, as he put it, 'the frequency of light-vibrations diminishes'. This is a form of redshift, although it is a property of the spacetime of de Sitter's universe, not a Doppler effect, and does not imply that the Universe is expanding. But de Sitter knew about Slipher's work and was one of the first astronomers to accept the idea that spiral nebulae were distant objects, far beyond the Milky Way. Einstein was completely baffled, and wrote to de Sitter that it 'does not make sense to me'. But worse – or, from a modern perspective, better – was to come, in the form of a variety of possible universes.

Over the next few years, several people dabbled with the equations of the general theory as they applied to a universe (or universes). But the person who took them by the scruff of the neck and shook them into the form that defined the whole subject of relativistic cosmology was a Russian, Alexander Friedmann.

Friedmann had been born in what was then (and is now, once again) St Petersburg, in 1888. His father was a ballet dancer and his mother studied the piano, but they married when she was sixteen and her husband was nineteen, which didn't give her much chance to establish a career. The marriage only lasted until 1896, when Alexander was eight. Friedmann's father remarried and Alexander grew up in his father's new family. In spite of this artistic background, he studied physics and was well up to date with the new developments in quantum theory and relativity in the first decade of the 20th century. He also married young, at the age of 23, the year after he graduated from St Petersburg State University, where he worked for the next couple of years. In 1913 Friedmann got a job as a meteorologist at the Main Physical Observatory in Pavlovsk. When war broke out, he volunteered as a technical expert, making weather observations for the Russian air force; this involved dangerous flights (as an observer/passenger) over enemy territory, on the Austrian front, and at least one crash landing. He was awarded the Russian St George Cross for valour as a result of the work. Friedmann survived the turmoil of the 1917 Revolution unscathed – he had been active in left-wing politics since his youth and supported the Revolution – and was appointed as professor at Perm State University. But he had to flee when the area was overrun by the Whites during the civil war that followed the revolution. Eventually, in 1920, he was able to settle in what was now Petrograd to carry out research in meteorology at the Academy of Sciences, and soon became responsible for all meteorological observations in the Soviet Union. But he died young, from typhoid* contracted during a visit to the Crimea, in 1925. (By then, St Petersburg/Petrograd had metamorphosed once again, into Leningrad.) This was just three years after the publication of his revolutionary ideas about cosmology.

Although he was a meteorologist by profession, Friedmann had kept up to date with developments in relativity theory, including the general theory, as best he could during the turbulent years following his graduation. It seems that he started thinking about the cosmological implications of the general theory in 1917, as soon as Einstein's key paper reached him. As this was a sideline from his

---

* At least, that is the official version. The always entertaining but not always reliable George Gamow, who was one of Friedmann's students, said that Friedmann actually died from pneumonia, contracted following a chill caught while making high-altitude meteorological observations from the open basket of a balloon. It is certainly a matter of record that Friedmann took part in a flight to an altitude of 7,400 metres in July 1925, two months before his death.

main work, and with the chaos surrounding him, it took several years to knock his ideas into shape and get them published. But when they were published, they were a tour de force. Even better, although based on mathematics, the key ideas of Friedmann's paper can be explained simply in ordinary language.

Friedmann's most important insight was that Einstein's equations do not describe a single, unique universe, but allow for the possibility of a whole variety of universes, or models. Einstein's static universe and de Sitter's empty universe, for example, both turn out to be members of the family of universes described by Friedmann. Some of these models turn out to look rather like the Universe in which we live; others do not. Once these ideas were fully appreciated (after the work of Hubble and Humason), the key cosmological quest became the investigation of which one of these models most closely matches the observations of the real world. Some of these models include a cosmological constant; others do not. Many of the most interesting ones (in the sense that they seem to offer a chance of a match to the real Universe) do not, although this was not obvious, of course, in 1922.

Leaving aside the more exotic variations on the theme, which are primarily of interest to mathematicians, Friedmann's models offered three alternative descriptions of the Universe, all of which expand naturally, without a lambda term. Crucially, as Friedman spelled out, the expansion is caused by space itself stretching, not by anything moving through space. In the first kind of universe, the expansion continues forever, although it slows down as time passes, because of the gravitational influence of all the matter it contains. For obvious reasons, this is called an 'open' universe. At the other extreme, the universe expands for a time, but eventually gravity overcomes the expansion and causes it to collapse back upon itself. This is a 'closed' universe. There are different kinds of open and closed universes, some expanding faster than others. But there is also one unique special case, a universe which sits exactly on the dividing line between being open or closed. This model expands forever, but more and more slowly, never quite coming to a halt. It is called a 'flat' universe, by analogy with the way the surface of a sphere, or the Earth, would appear to be completely flat if the sphere were expanded to a huge size. Without giving too much away of the story to come, it turns out that our Universe is indistinguishable from a flat universe, although it might be just open or just closed.

Friedmann wrote to Einstein summarising his work before the 1922 paper was published, seeking the great man's approval. Einstein's reply has been described by Gamow as 'a grumpy note', dismissing the idea. Friedmann published anyway, and Einstein then published a short paper (just eleven lines long!) saying that Friedmann's results were not compatible with his (Einstein's) equations. Then he had second thoughts and in 1923 published another note withdrawing these comments. It seems that Einstein then regarded Friedmann's solutions to the 'field equations' of the general theory as merely mathematical gimmicks, with no relevance to the real world. In a draft version of the 1923 note, preserved in the Einstein Archives, there are a few key words which were omitted from the published version. He said that 'a physical significance can hardly be ascribed' to Friedmann's models. Within a decade, he would be obliged to change his mind.

It might have happened sooner, if Friedmann had lived. In 1923 Friedmann presented his ideas in a book, *World as Space and Time*, in which he took the equations at face value, including the obvious implication that if the Universe is expanding it must have been smaller in the past, and very long ago it must have been very small indeed. He favoured the idea of a cyclic universe, which expanded from a very small state (perhaps a point), reached a certain size, then collapsed down into a point, before 'bouncing' into another cycle of expansion and collapse. He wrote:

> For example, it is possible that the radius of curvature constantly increases from a certain initial value; it is also possible that the radius changes periodically. In the latter case the Universe compresses into a point (into nothingness), then increases its radius to a certain value, and then again compresses into a point. [...] So far it is useless, due to the lack of reliable astronomical data, to cite any numbers that describe the life of our Universe. Yet if we compute, for the sake of curiosity, the time when the Universe was created from a point to its present state, i.e., time that has passed from the 'creation of the world,' then we get a number equal to tens of billions of usual years.[23]

This was published in 1923! This was the first scientific discussion of what became known as the Big Bang, and the first cosmological estimate of the age of the Universe (only a few times too big). But, like Einstein, the scientific world wasn't ready for Friedmann's breakthrough in the early 1920s, and with his early death there was nobody around to promote the cause – until another researcher independently came up with similar ideas.

A Priestly intercession

Georges Lemaître was six years younger than Friedmann, having been born in 1894 in Charleroi, in Belgium, and was educated at Jesuit schools. He was twenty in 1914, planning a career in civil engineering, but volunteered for military service in the First World War; his experiences in the conflict, where he was awarded the Belgian Croix de Guerre, made a deep impression on him, and contributed to his decision to develop parallel careers in science and the priesthood, although apparently he had expressed interest in becoming a priest at the tender age of nine. In 1920 he received a doctorate in physics from the University of Leuven, which in Belgium at the time was about the equivalent of a modern MSc, then studied

theology and was ordained in 1923. Alongside his theological studies, Lemaître prepared a thesis on relativity theory, which gained
him a studentship which he used to spend a year at the University
of Cambridge (1923–4) with Arthur Eddington. Eddington described
him as 'a very brilliant student, wonderfully quick and clear sighted,
and of great mathematical ability'.[24] From Cambridge, Lemaître
moved on to Harvard College Observatory, working with Harlow
Shapley for the academic year 1924–5, towards the end of the
debate about the nature of the spiral nebulae, and alongside (among
others) Cecilia Payne. While in America, he met Slipher, attended
the meeting in Washington DC where Hubble's measurement of
the distance to the Andromeda Nebula was announced, and visited
Hubble to find out more about his distance measurements to nebulae. Hubble's announcement fired his interest in applying the general
theory of relativity as a description of the real Universe. From the
outset he was interested in the physical significance of the redshift
measurements.

Lemaître's work in Harvard led to the award of a PhD; just
as Payne's doctorate was awarded by Radcliffe College, since the
Observatory did not award doctorates at that time, Lemaître's came
from MIT, in 1927, for a thesis on 'The Gravitational Field in a Fluid
Sphere of Uniform Invariant Density According to the Theory of
Relativity'. The relevant equations also apply, of course, to a universe of uniform density, although there things get interesting when
the density is not invariant but changes as time passes. Some of this
work was published in a paper in 1925, where Lemaître showed that
the radius of such a universe would be 'time-increasing' – that is,
the distances between all points in space would be growing. He was
the first person to attribute this to an actual expansion of spacetime.
But it made no impact. By the time this American doctorate was
awarded, Lemaître was back in Belgium, working at the University
of Leuven. There, he tackled more thoroughly the problem of

155

reconciling cosmological models based on the general theory of relativity with the redshifts revealed by Slipher's work.

This was the key difference between Lemaître's approach and that of previous researchers such as Friedmann (whose work he was unaware of at the time) and de Sitter. From the outset, rather than developing mathematical models for their own interest, he was trying to match theory and observation.

Lemaître was the first person to suggest that galaxies might be regarded as the equivalent of the 'test particles' in de Sitter's expanding universe, but he improved on de Sitter's work (and essentially independently reproduced Friedmann's results) by finding solutions to Einstein's equations in which the size of the universe (measured in terms of the distance between these test particles, or more technically in terms of a 'curvature parameter' sometimes referred to as 'the radius of the universe') varies in different ways. His preferred model was a closed universe in which this size changes as time passes, so that the size of the universe measured this way grows or shrinks. Because he knew of Slipher's work, he saw the expanding models as possible descriptions of the actual Universe, but he kept the cosmological constant as a parameter, which allowed for a greater choice of possible universes.

Slipher's evidence that the redshift of a galaxy seems to be bigger for fainter, and therefore presumably more distant, galaxies then led Lemaître to favour one particular kind of cosmological model, in which the 'velocity'* of a galaxy is proportional to its distance. This is what became known as 'Hubble's law'. It should be known as Lemaître's law, but it was originally published in 1927 in a Belgian journal that was not widely read outside that country, and through a series of accidents remained largely unknown until 1931.

---

* Actually a pseudo-velocity, since space is stretching rather than galaxies moving through space.

The title of the paper, though, should certainly have attracted attention in the relevant quarters. It translates as 'A Homogeneous Universe of Constant Mass and Increasing Radius Accounting for the Radial Velocities of Extra-Galactic Nebulae'. Lemaître had even sent a copy to Eddington, who must take most of the blame for failing to spread the word. The word deserved to be spread. A key passage in the paper read:

> When we use co-ordinates and a corresponding partition of space and time of such a kind as to preserve the homogeneity of the universe, the field is found to be no longer static; the universe becomes of the same form as that of Einstein, with a radius no longer constant but varying with the time according to a particular law.

That law is what became known as Hubble's law. Lemaître used Slipher's redshifts (interpreted as 'radial velocities') – as gathered together by Gustaf Strömberg in a paper published in 1926 – and distances based on a formula Hubble had derived relating the apparent brightness (magnitude) of a galaxy to its distance. This was a very rough and ready way of estimating distances, but was enough for Lemaître to determine the relationship between redshift and distance, coming up with a figure of 575 km per second per megaparsec for what is now known as Hubble's constant. This was after subtracting out the motion of the Milky Way discovered by Slipher. This is so close to the value published a couple of years later by Hubble that there is more than a suspicion that, as cosmologist Jim Peebles put it in his book *Modern Cosmology*, 'there must have been communication of some sort between the two'. Did the vain and pompous Hubble try to write Lemaître out of the story, as he also tried with Slipher? If so, it would be entirely in character.

Lemaître soon had an opportunity to discuss his work with Einstein, at a scientific meeting (a Solvay Congress) in the autumn of 1927. Thirty years later, he recalled in a radio interview that Einstein had described the model as 'abominable' from a physical point of view, whatever the equations said,* and seemed very ill-informed about astronomical observations, including Slipher's redshifts. Almost certainly, it was during these conversations with Einstein that Lemaître learned about Friedmann's work for the first time. A few months later, in 1928, at a meeting of the International Astronomical Union, de Sitter also gave the relatively unknown Belgian priest the brush-off.**

Undaunted (well, perhaps a little daunted), Lemaître continued to develop his ideas, and although he did not make vigorous efforts to promote them, on 31 January 1929, still before Hubble and Humason had published their first paper on redshifts and distances, he explained to a meeting in Brussels the crucial point that space itself is stretching as time passes to produce the redshift effect, and that this is not a Doppler effect caused by the motion of galaxies through space. As he had put it in the 1927 paper, the redshifts are 'a cosmical effect of the expansion of the universe'.

With all this out in the open, but being ignored, Lemaître was understandably somewhat taken aback when Hubble's work with Humason was promptly accepted and lauded a few months later. He wrote to Eddington, reminding him of the 1927 paper; one of Eddington's students, George McVittie, later recalled: 'I well remember the day when Eddington, rather shamefacedly, showed me a letter from Lemaître [...] Eddington confessed that, though he had seen Lemaître's paper in 1927, he had completely forgotten it

---

* 'Your calculations are correct, but your physical insight is abominable.'

** There is one intriguing possible reason why physicists largely ignored these exciting ideas. This was just at the time of the great breakthrough in quantum theory which revolutionised understanding of the subatomic world. With so much attention focussed on those developments, the general theory of relativity and cosmology were seen as rarified, exotic interests of no practical value.

until that moment.'[25] Eager to make amends, Eddington published a letter in *Nature* on 7 June 1930, drawing attention to Lemaître's work, and arranged for the translation of a slightly revised version of the 1927 paper (now including a mention of Friedmann, but curiously omitting his estimate of the 'Hubble constant') to be published in the *Monthly Notices of the Royal Astronomical Society* in 1931. But even before the English version of the paper was published, Lemaître's work became widely known, thanks to its endorsement by Eddington and by de Sitter, who learned about it from Eddington. From then on, Lemaître was recognised as a major player in the cosmological game, and he would be the one to take things to the next stage, introducing the idea of the Big Bang.

75

Sizing up the cosmic soufflé

In the early 1930s, the expanding universe idea became widely accepted following the work of Hubble and Humason (though not by Hubble himself, who preferred to make observations and measure distances rather than frame hypotheses). Even Einstein now accepted the evidence and, in April 1931, said as much at a meeting in Pasadena, California. But what was the Universe expanding from? In 1927, Lemaître had avoided any speculation about the origin of the Universe by suggesting that the observed expansion had started out from a static state, like Einstein's universe, which had been hovering on the edge of expansion for an indefinite time (perhaps forever) before the expansion started. This was one of the more esoteric possibilities allowed for by a judicious application of the cosmological constant. But by the time of Einstein's endorsement he was working on another idea.

In January 1931, Eddington gave a talk at a meeting of the British Mathematical Association, later published in *Nature*, in which he imagined running the expanding Universe backwards in time, so that galaxies got closer and closer together and eventually merged. The implication that there must have been a beginning to the Universe was, he said, 'utterly repugnant'. Later that year, Lemaître responded, also in *Nature* (in a paper with the splendid title 'The Beginning of the World from the Point of View of Quantum Theory'), that the beginning of the Universe 'is far enough from the present order of nature to be not at all repugnant', and speculated that 'we could conceive the beginning of the universe in the form of a unique atom, the atomic weight of which is the total mass of the universe. This highly unstable atom would divide in smaller and smaller atoms by a kind of super-radioactive process'. This was nothing more than a speculation, and Lemaître should have referred to a primeval atomic nucleus, rather than an atom. But at nuclear densities, winding back the entire observable Universe, the size of such a primeval entity would only have been about 30 times the diameter of the Sun; it would fit within the Earth's orbit. 'Naturally,' Lemaître said, 'too much importance must not be attached to this description of the primeval atom.' He acknowledged that it 'will have to be modified, perhaps, when our knowledge of atomic nuclei is more perfect'. But the key idea was that the Universe was born in violence – 'fireworks', as he put it – from a superdense state.

Lemaître did develop his ideas further, later hitting upon the term 'cosmic egg' to describe the superdense object out of which the Universe as we know it was born, culminating in his book *Hypothesis of the Primal Atom*, published in 1946. Lemaître's idea was a big influence on the work of Gamow and his team on what became known as the Big Bang (see Part Zero). But both in the 1930s and 1940s – and beyond – there was a major problem with the idea. The timescale was too short. Using the value for Hubble's constant found by Lemaître and by Hubble, the time since the outburst from the cosmic egg (or the Big Bang) would only have been about a billion years, far less than the by then well-established ages of the Sun and stars. Lemaître suggested that one way round this uncomfortable dilemma might be by a judicious use of the cosmological constant. According to the equations, it was possible for a universe to begin expanding from a superdense state, then to slow its expansion rate almost to zero, 'hovering' for as long or as short a time as you liked before the expansion began again.* But even in the 1930s this looked con-trived. Nevertheless, it is intriguing that Lemaître, who was always interested in the physical significance of the equations, always included a lambda term in his models, and regarded this as repre-senting a real physical entity, representing the vacuum energy of 'empty' space. Once again he was (eventually) to be proved cor-rect. But the version of the cosmic expansion idea which became established as the benchmark, and in which the conflict with stel-lar ages appeared with full force, emerged in 1932 and held sway for the rest of the 20th century. One of its authors was Einstein himself, but first, unknown to his colleagues, he had dabbled with a more radical idea.

Einstein's lost model

In 1931, shortly after the visit to Mount Wilson where he met Hubble, Einstein came up with the idea of a steady-state universe, which is infinitely old and eternally expanding, but in which new matter, or new galaxies, are constantly being created to fill the gaps between old galaxies as the space between them stretches. He got as far as drafting a paper in German (written on American paper), the title of which translates as 'On the Cosmological Problem', before deciding that there was a flaw in the argument and putting the work to one side. Although the document remained in the Einstein archives following his death, and has been readily available to schol-ars, it had been misfiled as an early draft of another paper with a similar title. Nobody read it for decades – or if they did, they missed its importance. It was only in 2013 that its significance was noticed

---

* As this example highlights, if you are allowed to choose any cosmological constant you like, you can get a universe that does just about anything you want in the way of expanding and shrinking (or hovering). This makes the constant useless when it comes to predicting what the Universe 'ought' to be like.

by Cormac O'Raifeartaigh and Brendan McCann, of Waterford Institute of Technology, leading to its publication in English in 2014.

At the beginning of 1931, Einstein was a recent convert to the idea of an expanding Universe and, still unhappy with the idea of a Universe that changes as time passes, was looking for a way to reconcile this with his conviction that the Universe has always looked the same as it does today. The steady-state idea does this neatly; the universe observed from any galaxy always looks the same *on average*, even though individual galaxies fade from view as time passes and are replaced by new ones. You might make a rough analogy with a very old wood (or a tropical rainforest). The wood has been there for thousands of years, but over that time many generations of trees have grown up, died, fallen and been replaced by younger trees. The idea itself is fairly obvious, but Einstein, being Einstein, wanted to frame it in the mathematics of the general theory of relativity.

He did so with a reinterpretation of the cosmological constant, which he no longer needed in order to stop space expanding. Or as he put it, 'this solution is [now] almost certainly ruled out for the theoretical comprehension of space as it really is'. Misspelling Hubble's name (as he did consistently in his writings at that time), he went on:

> Hubbel's exceedingly important investigations have shown that the extragalactic nebulae have the following two properties:
> 1) Within the bounds of observational accuracy they are uniformly distributed in space
> 2) They possess a Doppler effect* proportional to their distance

Einstein suggested that the expansion of the universe is driven by the creation of the new matter required to keep the overall density

---

* Surprisingly careless; if anyone should have appreciated that this is *not* a Doppler effect, it should have been Einstein!

of the universe constant as it expands. He wrote:

> If one considers a physically bounded volume, particles of matter will be continually leaving it. For the density to remain constant, new particles of matter must be continually formed in the volume from space.

This is uncannily like the idea of a 'creation field', or C-field, developed by Fred Hoyle, in complete ignorance of Einstein's unpublished work, from the late 1940s onward. But unlike Hoyle, instead of invoking a separate C-field, Einstein identified the creation process with the cosmological constant. At which point, as he soon realised, his argument broke down. The natural solution to the equations with this use of the cosmological constant is that space is empty (zero density) and there is therefore no creation of matter! Einstein's handwritten corrections show that he spotted this point, but modern readers are left baffled that he did not introduce a separate creation field, like Hoyle's C-field. Most probably, it was because of his conviction that the Universe must be simple (remember that he later referred to the introduction of the cosmological constant as the 'biggest blunder' of his career). This love of simplicity soon showed up in another cosmological model, which he developed with the Dutch astronomer Willem de Sitter, and published in 1932.

Keeping it simple

It was also in 1932 that James Jeans, a British physicist who was also a great populariser of science, wrote:

> On the face of it [it] looks as though the whole universe were uniformly expanding, like the surface of a balloon while it is being inflated, with a speed that doubles its size every 1,400 million years [...] If Einstein's relativity cosmology is sound, the nebulae have

no alternative – the properties of the space in which they exist
compel them to scatter.

This highlights the difference between the joint Einstein–de Sitter
model and their individual, pre-Hubble (and pre-Lemaître) ideas.
One of the key features of the model developed by Einstein and
de Sitter together is that from the outset it was motivated by the
links with observation – unlike the earlier work of Friedmann, de
Sitter himself or even Einstein, which was essentially motivated by
an interest in the mathematics of the general theory. The paper by
Einstein and de Sitter, drafted in January 1932 and published in March
that year, had the title 'On the Relation between the Expansion and
the Mean Density of the Universe'. It is only two pages long, and in
a sense it doesn't say anything about cosmological models that had
not already been said by Friedmann and Lemaître. For this reason,
it has sometimes been suggested that it only got published at all, let
alone taken notice of, because of Einstein's name. But that is wrong.
The importance of the paper is that it attempts to describe the real
Universe, not an abstract mathematical model. The title refers to
'the Universe', not to 'a universe'. This was a crucial leap.

Einstein and de Sitter knew that the Universe was expanding,
and they had a number (Hubble's constant) for the speed with which
it was expanding, even though we now know that this number was
rather too big. The other key feature of the Universe which ought to
be susceptible to observation, they realised, was its density, measured
in terms of counting the number of galaxies in a chosen volume of
space, and calculating the equivalent average density if all the matter in all the stars in those galaxies was spread out evenly through
space. These two numbers together would be enough to determine
the ultimate fate of the Universe – whether it was expanding fast
enough to escape from its own gravitational pull and expand forever
('open', with 'negative curvature'), or whether the density was so

high that gravity would eventually first halt the expansion and then cause the Universe to collapse back into a superdense state ('closed', with 'positive curvature'). But there is just one special case, arguably the simplest solution to the equations; this is what attracted the attention of Einstein and de Sitter.

It is possible to describe mathematically (and very simply), using the equations of the general theory, a universe which sits exactly on the dividing line between being open and closed – the so-called 'flat' universe. In the simplest possible version of this, the universe is also homogeneous (the same everywhere) and isotropic (the same in all directions). Friedmann, as we have seen, 'discovered' the flat universe model, along with all the other mathematical possibilities. But he did not relate this to observations of the real Universe. Nor did Lemaître. That is what made the Einstein–de Sitter paper of 1932 special. They pointed out that if the value of Hubble's constant was known, then the density required to make the Universe flat could be calculated and compared with observations of the Universe. With Hubble's value for the constant (500 km/second per megaparsec), the required density is 4×10^{-28} grams of matter in every cubic centimetre of space. Because modern estimates of the Hubble constant are (for reasons that will soon become clear) nearly ten times smaller than Hubble's value, the modern version of this calculation suggests a lower density, a little more than 10^{-29} grams per cubic centimetre. If all of this matter were in the form of hydrogen atoms and spread out evenly, this density would correspond to a single atom of hydrogen in every million cubic centimetres of space.

It's worth mentioning that this was, at one time, a powerful argument in favour of the steady-state idea. You only need to make a very few atoms of hydrogen, dotted across the universe, to fill in the gaps as the universe expands. As Fred Hoyle used to point out, this is no more objectionable, in principle, than the idea that all the matter in the Universe was made at once, in a Big Bang. Hoyle is sometimes presented today as a crank with a crazy idea, but (as the fact that Einstein came up with the same idea highlights) the steady-state model was at the time (and right through to the discovery of the cosmic microwave background radiation) a respectable alternative to the Big Bang model.

Even in the 1930s, it was clear that there was not enough matter in all the bright stars in the visible galaxies to make the Universe flat. But it was also clear that there was nearly enough matter to do the job – 'nearly', that is, when you considered the vast range of mathematical possibilities. Rather than talking in terms of the number of hydrogen atoms per million cubic centimetres, cosmologists use a number called the density parameter, usually denoted by the Greek letter omega (Ω) which is defined so that in a flat universe $\Omega = 1$. If the Universe contains half the matter required to make it flat, $\Omega = 0.5$; 30 per cent of the required matter corresponds to $\Omega = 0.3$; and so on. Roughly speaking, the amount of visible matter in our Universe corresponds to about $\Omega = 0.1$. In other words, there is about a tenth (actually very slightly less) as much visible matter as required to make the Universe flat. But the equations allow for the possibility of universes with any value for Ω – a billion, or a billionth; a trillion trillion, or a trillionth of a trillionth; or any other number. Even in the early 1930s, when cosmology first became a quantitative science, it was clear that the density of the real Universe was suspiciously close to the value required to make the Universe flat.[26] It seemed natural to Einstein and de Sitter to suggest that the Universe really is flat, but that we can't see everything it contains. Although the estimated density of the Universe in 1932 didn't quite match the model, they said that:

> It certainly is of the correct order of magnitude, and we must
> conclude that at present time it is possible to represent the facts
> without assuming a curvature of three-dimensional space. The

curvature is, however, essentially determinable, and an increase
in the precision of the data derived from observations will enable
us in the future to fix its sign and to determine the value.

All that is required to make $\Omega = 1$ is the presence of sufficient
unseen stuff – what we now call 'dark matter' – in addition to the
bright stars. Although the idea of there being enough dark matter
to flatten the Universe was not taken entirely seriously at the time,
there is another way to reconcile the observations with the idea
of a flat Universe: that is to improve measurements of the Hubble
constant, in the hope of finding that Hubble had overestimated it. A
small enough value for this number would allow the Universe to be
flat even with a lower density (and it would also increase estimates
of the time since the Big Bang, perhaps reconciling estimates of
the age of the Universe with estimates of the ages of stars). So the
Einstein–de Sitter model (flat, homogeneous and isotropic) became
the cornerstone of cosmology (partly because it was the simplest
model to work with), taught to generations of students (including
myself), in spite of the questions about the value of Ω and the
Hubble constant.* Over the decades that followed, the efforts of
cosmologists focussed on finding the value of the Hubble constant,
because that was all that could be done at the time. Then, as we shall
see, it became possible to measure the amount of dark matter in the
Universe, pinning down the value of Ω precisely.

Across the Universe
Doubling the distances
Part of the appeal of the Einstein–de Sitter cosmology is that it provides a very simple way to calculate the age of the Universe from

---

* De Sitter didn't live to see this. He died of pneumonia in London in November
1934, at the age of 62.

the Hubble constant, H. If the Universe had been expanding at the same steady rate since the Big Bang, then the age of the Universe (the time since the Big Bang) would be just 1 divided by H, once the 'kilometres' and the 'megaparsecs', both units of distance, are divided up to leave seconds, which are then converted into years. The resulting 'age' is known as the 'Hubble time'. But the expansion of the Universe has been slowing down since the Big Bang, so the value of H has decreased as time passed. The Hubble constant is 'constant' in the sense that it is the same everywhere in the Universe at a particular time (known as a cosmic epoch), but it changes as time passes. So astronomers sometimes refer to the Hubble 'parameter', and say that the Hubble constant is the value of the Hubble parameter at the present epoch. Because the Universe was expanding *faster* in the past, it has taken *less* than the Hubble time to get to its present state. But how much less? This is where the simplicity of the Einstein–de Sitter model becomes useful.

In the Einstein–de Sitter cosmological model, the age of the Universe is just two-thirds of the Hubble time. This is what gives us an age of a bit more than a billion years for a value of H of 500 km per second per Mpc, in clear conflict with the age of the Earth already known in the 1930s (and , in slightly less clear conflict, with the ages of the stars).

There is something else odd about a high value of the constant, apart from the conflict between the implied age of the Universe and the ages of stars, which very few people took notice of even in the 1930s. The way to determine the value of the Hubble constant was to measure distances to galaxies accurately, and compare these measurements with measurements of their redshifts. But once you have a value for the constant, you can use it to determine distances – which is why Hubble was interested in redshift measurements in the first place. You can use the Hubble constant as a measure of the distance scale of the Universe. The smaller the distances measured to other galaxies, the larger the value of the Hubble constant, because it has taken less time for cosmic expansion to have carried the galaxies to their present distances. Equally, turning that around, the bigger the constant, the smaller the distances between galaxies. Hubble's measurements, starting with Cepheids and working out into the Universe, clearly showed that the spiral nebulae are galaxies beyond the Milky Way, not star clouds within the Milky Way. But they were still, according to his measurements, relatively close to us and to each other, which meant, comparing their apparent sizes in the sky to their distances, that they must be considerably smaller than the Milky Way Galaxy. Could it be possible that we live in the biggest galaxy in the Universe?

Such a notion was not entirely implausible at the beginning of the 1930s, but one person who doubted it was Arthur Eddington. In his book *The Expanding Universe*, published in 1933 and one of the first popular accounts of the new discoveries, he commented:

> The lesson of humility has so often been brought home to us in astronomy that we almost automatically adopt the view that our own galaxy is not specially distinguished – not more important in the scheme of nature than the millions of other island galaxies. But astronomical observation scarcely seems to bear this out. According to the present measurements the spiral nebulae, though bearing a general resemblance to our Milky Way system, are distinctly smaller. It has been said that if the spiral nebulae are islands, our own galaxy is a continent. I suppose that my humility has become a middle-class pride, for I rather dislike the imputation that we belong to the aristocracy of the universe. The earth is a middle-class planet, not a giant like Jupiter, nor yet one of the smaller vermin like the minor planets. The sun is a middling sort of star, not a giant like Capella but well above the lowest classes. So it seems wrong that we should happen to belong to

an altogether exceptional galaxy. Frankly I do not believe it; it would be too much of a coincidence. I think that this relation of the Milky Way to the other galaxies is a subject on which more light will be thrown by further observational research, and that ultimately we shall find that there are many galaxies of a size equal to and surpassing our own.

This is an example of what has become known as the 'principle of terrestrial mediocrity', which holds that there is nothing special about our place in the Universe. Eddington was way ahead of his time, and little notice was taken of his remarks on this subject in the 1930s. But if you assume that the Milky Way is an average-sized spiral, and adjust the distance scale (by adjusting Hubble's constant) to move other spiral galaxies far enough away to make their calculated sizes the same (on average) as the size of the Milky Way, you have to reduce the value of the Hubble constant by roughly a factor of ten, and therefore increase the calculated age of the Universe from a bit more than 1 billion years to a bit more than 10 billion years. Eddington, though, stopped short of making this leap.* At the time he was writing, another way to satisfy the principle of terrestrial mediocrity would have been to find other galaxies the size of the Milky Way, beyond the range of the telescopes available in the 1930s, each surrounded by smaller spirals. It is only hindsight, based on surveys of vast numbers of galaxies, that makes it obvious which resolution of the puzzle applies in the Universe we live in. The Hubble–Lemaître measurement of the constant really was far too big. The first, and most dramatic, change in the accepted value of the Hubble constant was indeed a result of further observational

---

* Eddington's speculation was confirmed many years later. See End Note 21, p. 235.

research; it came in the 1940s, around the time that Gamow was enthusiastically promoting the Big Bang idea.

The breakthrough came about partly as a result of a combination of the Second World War and the vagueness of a German-born astronomer, Walter Baade, when it came to dealing with anything except astronomy. Baade had been born at Schröttinghausen in 1893, so he was just four years younger than Hubble. He completed his PhD at Göttingen in 1919, and worked at the Bergedorf Observatory of the University of Hamburg for eleven years, before moving to the United States in order to use larger telescopes than anything available in Europe. He became a member of the observing staff at Mount Wilson soon after Hubble and Humason had published the first of their papers on the redshift–distance relation. He worked with Hubble and others on studies of supernovae and distances to other galaxies, gaining a reputation as a fine observer. His personal life, however, was less well organised than his observing, and although he intended to become an American citizen and got as far as starting the application process in 1939, he lost the relevant papers during a house move, and the application lapsed. He still hadn't got around to renewing the application when the Japanese attacked Pearl Harbor in December 1941 and Germany, Japan's ally, declared war on the United States. This made Baade technically an 'enemy alien', and at first he was put under a curfew that required him to stay at home each night between 8pm and 6am, bringing his observing to a halt.*

Over the next few months, many of the astronomers (including Hubble) were recruited for war work or into the military, until Baade was the most senior astronomer left at Mount Wilson. It was eventually decided that he posed no threat to the USA, but equally

---

* He (like many other German-Americans) was luckier than many Japanese-Americans, who were very badly treated and put in internment camps.

he was not regarded as a suitable person to be involved in the war effort, so he was allowed to resume observing with the 100-inch telescope, just at the time when a new, more sensitive kind of photographic plate was becoming available and a night-time blackout was imposed on the city of Los Angeles. He had the best telescope in the world at his disposal, with the best possible plates and dark skies. This did not make studying faint stars in galaxies beyond the Milky Way easy, just slightly less difficult, but by 1943 Baade was a highly skilled observer. So he was able to photograph fainter objects than anything Hubble had been able to, and set about making a detailed survey of the Andromeda Galaxy.

Baade was able to pick out individual stars not only in the outer parts of the Andromeda Galaxy (where Hubble had found Cepheids) but also in the inner regions, which had previously appeared only as a blur on photographic plates. His first major discovery was that the Andromeda Galaxy is made up of two different kinds of star – with the implication that all spirals, including the Milky Way, have a similar structure. One kind of star, which Baade called Population I, is found in the outer part of the galaxy: the disc and spiral arms. These are hot, young stars, blue or yellow in colour and containing large amounts of heavy elements. Stars in the central region, the bulging nucleus of the galaxy, which he called Population II, are older, cool, red stars with very little in the way of heavy elements. The same kind of stars are found in globular clusters. Further investigations showed – as discussed in Part One – that Population II stars formed from the primordial material left over from the events immediately following the Big Bang, while Population I stars are relatively young, and formed from material that had already been processed in previous generations of stars. The same pattern does indeed apply to all spiral galaxies, and the Sun, relatively rich in heavy elements, is, of course, a Population I star.

Baade also found, in 1944, that there are two types of Cepheid variable, one associated with each stellar population. Population I Cepheids are now known as 'classical' Cepheids, and Population II Cepheids are often known as W Virginis stars, after the archetypal member of the group. Each kind of Cepheid has a characteristic period–luminosity relation, but overall the W Virginis stars are fainter than the classical Cepheids. In 1944, this discovery did not alter astronomers' understanding of the cosmological distance scale, because Hubble had used classical Cepheids, like the ones in our part of the Milky Way, in his work; there didn't seem to be any confusion. But once again new technology came along, leading to new discoveries that changed the astronomers' picture of the Universe.

.This time, the new technology was an even bigger and better telescope, the 200-inch reflector on Mount Palomar,* which became operational in 1948 and was the most powerful telescope available to astronomers on Earth for the next 45 years; it is still operating and doing valuable work. Transferring everything he had learned, and the best photographic technology, to the 200-inch, Baade confidently set out to study stars known as RR Lyrae variables in the Andromeda Galaxy. RR Lyrae stars are fainter than Cepheids, but they are very good distance indicators; they are often found in globular clusters, and Baade was sure he would be able to find them in the Andromeda Galaxy. But he couldn't. He could just pick out the brightest stars in the globular clusters, but not the fainter RR Lyrae stars. Fortunately, from studies of globular clusters in our own Galaxy, astronomers already knew how much brighter the brightest Population II stars in these clusters, red giants, are than RR Lyrae stars. If the red giants Baade was observing in the globular clusters in Andromeda had the same characteristics as those in the globular clusters in our Galaxy,

---

* The telescopes on Mount Wilson and Mount Palomar were all part of a single outfit, the Mount Wilson and Palomar Observatories, linked with Caltech.

the RR Lyrae stars would indeed be too faint to be detected with the tools he had available. But, in order to appear as faint as they did, those red giants had to be a lot farther away than the distance Hubble had determined for the Andromeda Galaxy. The reason for the error soon became clear – and it went right back to the original determination of the Cepheid distance scale by Shapley, 30 years before.

Shapley had used data for every Cepheid he could lay his hands on when he worked out the distance relationship. Unfortunately, this involved a mixture of what were known by the late 1940s to be two kinds of Cepheid, Population I and Population II. The Population I Cepheids are brighter, which you might think would make the mistake obvious. But they lie in the plane of the Milky Way (the disc) where there is a lot of dust (more dust than was realised in Shapley's day), and the dust dims the light from them. The population II Cepheids lie above and below the plane of the galactic disc, where there is less dust. By sheer bad luck, the dimming effect almost exactly cancels out the extra brightness of the Population I Cepheids used by Shapley. So, Hubble had been looking at Population I (classical) Cepheids in the Andromeda Galaxy, but had been using the distance relationship that actually applied to Population II Cepheids (W Virginis stars). The Cepheids used in his calculations were all brighter than he had realised; in order to appear as faint as they do, they had to be farther away. It meant that the Andromeda Galaxy was roughly twice as far away as he had calculated, and that the distance scale of the entire Universe had to be adjusted accordingly, bringing the value of the Hubble constant down to about 250 km/sec per Mpc. The result, announced in 1952, was a scientific sensation that even made newspaper headlines, with stories that the size of the Universe had doubled. But what was even more important was that the calculated age of the Universe had doubled, pushing it close to 4 billion years, not hopelessly different from the already known age of the Earth. Even in 1952, the ages of stars were not very well known, and 5 billion years seemed a reasonable estimate, so there was not too much concern at the remaining discrepancy. But as the 1950s progressed, although estimates of the age of the Universe increased farther, estimates of the ages of stars increased even faster, keeping the steady-state alternative to the Big Bang idea very much alive.

Hubble's heir

The increasing estimates of the age of the Universe in the 1950s – and on into the 1960s – were based on improved determinations of the Hubble constant. The man who did most to improve those determinations was another American, Allan Sandage, who became Hubble's scientific heir, and pushed the 200-inch telescope to its limits.

Sandage is the first person in this story who grew up knowing that the Universe is expanding. He was born in Iowa City in 1926, just a year before Lemaître published the redshift–distance relationship, and three years before this became known as Hubble's law. He 'discovered' astronomy at the age of nine, when he looked at the night sky through a school friend's telescope, and he read Hubble's book *The Realm of the Nebulae* and Eddington's *The Expanding Universe* when he was a teenager. Although his education was interrupted when he was drafted into the US Navy in 1944, on his release in 1945 he studied at the University of Illinois, graduating in 1948, and then at Caltech, where he worked for a PhD. His interest in cosmology was sparked by Fred Hoyle, who taught a course at Caltech as a visitor while Sandage was a student there. Sandage completed his PhD (supervised by Walter Baade) in 1953, the year Baade 'doubled the size of the Universe'; by this point he had started working at the Mount Wilson and Palomar Observatories on a project devised by Edwin Hubble, one of his boyhood heroes. He worked there for the rest of his career.

The project Sandage was recruited to work on was an attempt to measure the flatness of the Universe – how closely the real Universe matches the Einstein–de Sitter model. This was effectively the equivalent in three dimensions of measuring the flatness of a two-dimensional surface, like a sheet of paper. On a flat surface, as we were taught in school, the angles of a triangle add up to 180 degrees, and if we know the lengths of the sides of the triangle we can work out its area. On the surface of a sphere (closed), the angles of a triangle add up to more than 180 degrees, and the area of the triangle is correspondingly bigger; on a surface curved like a saddle or a mountain pass (open), the angles of a triangle add up to less than 180 degrees, and the area is correspondingly smaller.

Translating this into three dimensions involves measuring volumes rather than areas. If space is curved (one way or another), the number of galaxies counted at different distances from us will be different from the number counted if space is flat. Sandage was set the task of counting galaxies, using photographic plates obtained with a wide-angle telescope called a Schmidt camera. The Schmidt could photograph a wide area of sky in one plate, whereas the 200-inch could see farther out but only in a tiny field of view. The Schmidt plates did not contain redshift information, but as a first step Hubble thought (correctly) that it would be reasonable to assume that fainter galaxies are farther away. The counting and collating was just the kind of job to give to a graduate student – tedious, painstaking work at the end of which you might get an acknowledgement at the end of a scientific paper. Not for nothing is the work known as 'number counting'.

Sandage didn't even get to go up the mountain to do any observing at first. Then, in the summer of 1949, Hubble suffered a heart attack, and his doctor insisted that he did not go up the mountain while he was recuperating. Sandage and another student, Halton

Arp, were sent to learn observing from Walter Baade, since it was clear that if Hubble ever did return to work he would need help.*
The project they cut their teeth on involved photographing and analysing globular clusters, initially using a 60-inch telescope, then, having proved themselves as observers, moving on to the 100-inch. This was the work for which Sandage was awarded his PhD, having turned out to be a first-rate observer. In 1952, he made the first studies of the 'turnoff' technique which, as explained in Chapter Four, is a key to measuring the ages of globular clusters.

By then, though, Sandage was already working as Hubble's assistant. Hubble had planned out a concentrated attack on the distance scale, building from Baade's discovery to pin down the value of Hubble's constant more accurately and – although Hubble did not think of it in those terms – thereby to determine the age of the Universe. Sandage was now making the observations with the 200-inch that Hubble had planned to make himself, but which he could no longer carry out, even though he was allowed to make limited visits to the mountain from October 1950. The situation was formalised into a paid appointment as Assistant Astronomer in 1952, but Sandage then spent a year as a visitor at Princeton, following up the implications of the discovery of the main-sequence turnoff. He intended to develop this further and study stellar evolution. Shortly after he returned to Caltech, however, in September 1953 Hubble died as a result of a stroke, just short of his 64th birthday. Humason and Baade were also both now in their sixties. It was time for the next generation to pick up the reins, and Sandage was the leading observer of the new generation. Somewhat reluctantly, out of a sense of duty, he took up the task:

---

* Incidentally, in case you are wondering, 'number counts' show that the Universe is indeed flat – or, as the experts more cautiously put it, that there is no evidence for curvature.

I felt a tremendous responsibility to carry on with the distance-scale work. [Hubble] had started that, and I was the observer and I knew every step of the process that he had laid out. It was clear that to exploit Walter Baade's discovery of the distance-scale error, it was going to take 15 or 20 years, and I knew at the time it was going to take that long. So, I said to myself, 'This is what I have to do.' If it wasn't me, it wasn't going to get done at that period of time. There was no other telescope; there were only 12 people using it, and none of them had been involved with this project. So I had to do it as a matter of responsibility.[27]

He was just 27 years old.

The project Sandage embarked on was a complete revision of the cosmological distance scale determined by Hubble, starting out with more detailed observations of more Cepheids. Like Hubble's original work with the 100-inch and following Hubble's plan for the 200-inch, after Sandage had determined the distances to nearby galaxies using the Cepheid technique, he could identify brighter objects in those galaxies and calibrate their brightness by comparison with Cepheids, so that they could in turn be used as 'standard candles' to indicate the distances to more remote galaxies. At this stage in the study, it turned out that Hubble had made an understandable mistake. He had used what he thought were very bright stars in other galaxies as standard candles; the greater power of the 200-inch showed that these were not bright stars, but glowing clouds of gas known as HII regions. HII regions are also known in our own Milky Way Galaxy, so their brightness could be calibrated. It turns out that there is a maximum brightness they can have; so, identifying the brightest HII regions in a galaxy and measuring their apparent brightness is a good guide to distance. But HII regions are brighter than the stars Hubble had compared them with, which means the galaxies were farther away than he had thought. As with

Baade's correction to the distance scale, this implied a smaller value for the Hubble constant, a further reduction in addition to Baade's reduction.

Sandage's first contribution was made using redshift and brightness data for 850 galaxies, compiled by Milton Humason and a younger colleague, Nick Mayall, over the previous two decades. The three of them published a joint paper in 1956 which confirmed that the Lemaître–Hubble law (redshift is proportional to distance) holds out as far as they could measure it, to redshifts corresponding to 'velocities' of 100,000 km per second, one-third of the speed of light. Overall, including Baade's correction and combining these measurements and the HII region evidence, this step alone meant that galaxies are three times farther away than Hubble had thought, and that the value of the Hubble constant is no more than 180. But this was just a first step. This would be a recurring theme of Sandage's work in the 1950s and beyond: every improvement he made using the 200-inch reduced the value of the constant. As the years passed, the reductions added up (or subtracted down) more and more.

Globular clusters can also be used as standard candles, because once distances to enough nearby galaxies had been measured it became clear that the brightest globular clusters all have much the same brightness, whatever galaxy they are in. As Sandage gradually and painstakingly built up his database, he found that even whole galaxies can be used as distance indicators, because in a large cluster of galaxies there is usually one very bright galaxy, and the brightest galaxies in different clusters all have about the same intrinsic brightness.

The key step in all this work was a measurement of the distance to a large cluster of galaxies, which lies in the direction of (but far beyond) the constellation Virgo, and is therefore known as the Virgo Cluster. Our own Milky Way, its companions the Magellanic Clouds and the Andromeda Galaxy are all part of a small group of galaxies (the Local Group) bound together by gravity, just as the stars of the whole Milky Way are bound together in a single system by gravity. Although measuring distances across the Local Group is good for things like calibrating the brightness of other objects compared with Cepheids, these measurements can tell us nothing about the red-shift–distance relation – the Andromeda Galaxy is actually moving *towards* us, so its light shows a blueshift. On such a scale, gravity overcomes the stretching of space. The full effect of expanding space is only seen between clusters of galaxies (or between clusters and our Local Group), which are the equivalent of spots (what cosmologists call 'test particles') on the surface of James Jeans' imaginary expanding balloon, being carried apart as the surface of the balloon stretches. The Virgo Cluster contains more than 2,500 galaxies, full of things like globular clusters that can be used as distance indicators. Once the distance to the cluster was known, Sandage could make bigger leaps out into the cosmos. With the 200-inch, Cepheids took Sandage out to a distance of 5 million light years; HII regions took him out to tens of millions of light years; the Virgo Cluster to about 65 million light years away. Sandage eventually pushed his distance estimates out beyond 300 million light years, using galaxies as standard candles; this was far enough to be sure that he really was sampling enough of the Universe to draw conclusions about the redshift–distance relationship.

By 1958, the conclusion he had drawn was that the value of the Hubble constant is about 75, in the usual units, but – allowing for the uncertainties involved in all the steps in the investigation – it might be a little less than 50 or a little more than 100. It was, however, a long time before this figure became widely accepted.

The problem was that there was no consensus about the Hubble constant. Other astronomers using different techniques and making different allowances for things like interstellar extinction came up with their own numbers, and these were all considerably higher than Sandage's number. Sandage was the only person to apply all of the corrections. But at the beginning of the 1960s, there were at least three other seemingly respectable estimates – one placed the number in the range from 143 to 227, another between 120 and 130, and a third in the 130s. Even though it was widely accepted that Sandage was an expert at this job, and that the 200-inch was the best telescope for the job, the fact that there were other opinions biased the thinking of the astronomical community towards the top end of the range that Hubble had considered. By the time I began studying astronomy seriously in the mid-1960s,* the number generally used by cosmologists was 100 km/sec per Mpc, with the view that this might be a little on the high side but that it was a nice round number to work with.

There were two problems with this. The first problem everybody knew but ignored: $H = 100$ gives an age of the Universe of less than 9 billion years, and the ages of globular clusters were then being estimated as in the mid-teens of billions of years, with some uncertainty but not enough to take the ages below 10 billion years. The other problem nobody seemed to know or care about. As a student I had read Eddington's comments about the size of the Milky Way and been suitably impressed. With $H = 100$, the Milky Way would be about twice as big as other spirals. Less dramatically, but still a matter of concern to the younger me, if H were bigger than 70, the Milky Way and the Andromeda Galaxy would each be bigger than any of the thousands of galaxies in the Virgo Cluster. This puzzled me. But I was in no position to argue with my superiors, who, when I mentioned this, more or less patted me on the head and told me not to worry about it, to leave it to the grown-ups. This highlights a serious point. In the early 1960s, nobody (except possibly George

---

* My less serious interest dated back to George Gamow's books, which I read in the mid-1950s.

Gamow and Georges Lemaître, who were still alive) really believed, deep down, that there had been a Big Bang. Cosmology was still a kind of academic game, played only by a handful of people using equations, and it didn't really matter if the equations didn't quite match observations of the real world.

It was the age of the Universe problem, of course, which made the steady-state alternative to the Big Bang respectable through the 1950s and into the 1960s, until Penzias and Wilson came along and pulled the cosmic microwave background out of the hat. In doing so, they made cosmology more than a game and simultaneously made the Big Bang the best bet and the age of the Universe problem something to take seriously. But, before moving on to modern measurements of the Hubble constant and the age of the Universe, there is much confusion about exactly what the steady-state model was, and it is worth a little digression to clear up the confusion.

Another Great Debate

In 1947, the Royal Astronomical Society asked Hermann Bondi, a young Austrian-born researcher then working at the University of Cambridge, to write a review summarising everything that was known about cosmology. This turned out to be an enormously influential article, which stimulated the development of cosmological thinking in Britain.* Bondi covered all the ideas discussed here (and more!), and drew attention to the power of the general theory of relativity in cosmology. He also highlighted the critical problem of the time. 'Some models of the universe assume a catastrophic origin,' he said, 'meanwhile, other theorists are more conservative, and do not embrace the concept of an explosive origin for the universe.'

---

* Some twenty years later, when I had ambitions to become a cosmologist, Bondi advised me to do something sound first before venturing into such speculative territory; so I did my PhD work on stars. I've never been sure this was sound advice.

Notice that in the late 1940s the *conservative* view was that there had *not* been an explosive origin. While preparing the article, Bondi discussed the ideas with Hoyle and their contemporary Tommy Gold, another Austrian-born astronomer-physicist, who eventually moved to the United States. Bondi and Hoyle in particular worried about the fact that the mathematical solutions to Einstein's equations developed in the 1920s and 1930s did not offer an explanation for the existence of matter, except for Lemaître's fireworks, which they found philosophically unsatisfying. But it was Gold who came up with the bright idea that set them on the track leading to the steady-state model.

One evening the three friends went to the cinema to see the movie *Dead of Night*, a horror story built around a recurring nightmare. The story has no beginning and no end: you could start watching at any point and sit there until you got round to the same place, and you would always get the same cinematic experience. It was Gold, a few days later, who suggested the idea that the Universe might be like that. You could 'join' it at any time, and it would always look the same. Perhaps it had no beginning and no end. This gave them an alternative to the cosmic fireworks of Lemaître: an expanding universe that always looked the same on average, with new matter being created in the void between galaxies as they moved apart – continuous creation rather than all-at-once creation.* Although at first repelled** by the idea of continuous creation, they soon convinced themselves that this was no worse

---

* Gold was always an 'ideas man' and good at getting them noticed. In the 1960s, when pulsars were discovered, there was much talk at the Institute of Theoretical Astronomy in Cambridge (where I was busy *not* being a cosmologist) about possible explanations, including the idea that they might be spinning neutron stars. As I recall, the idea emerged in group discussions over coffee. But it was Gold who quickly dashed off a paper to *Nature* and got credit for the suggestion.
** Hoyle's word.

than the alternative, since, after all, matter had to have come from somewhere.

At first, the three of them planned to write a joint scientific paper about their idea. But it quickly turned out that they had different views about how to go about this. Bondi and Gold were more interested in the – for want of a better word – philosophical aspects of the model, while Hoyle was excited by the prospect of tying it into the framework of the general theory of relativity. He did this by introducing the C-field (C for creation) – discussed at the start of this chapter – and linking this to the expansion of the universe. In his memoir *Home is Where the Wind Blows*, Hoyle elegantly explains that in order to compensate for the positive energy of the newly created particles, the C-field carries negative energy out into the Universe, and this causes the expansion. According to Hoyle, this fell naturally out of the equations and 'greatly surprised' him. The steady-state universe expands *because* there is continual creation. Overall, energy is in balance with no net loss or gain. In 1948, he argued that the new particles that had to be created in expanding empty space would probably be neutrons, since neutrons spontaneously decay into protons and electrons, the constituent parts of hydrogen atoms – at a rate of one hydrogen atom per cubic metre every 10 billion years. This, in turn, would be one of the things that led him to investigate the way other elements are synthesised inside stars. So two separate papers, one by Bondi and Gold, the other by Hoyle, were published in 1948.

A flavour of the Bondi–Gold approach can be gleaned from a term they coined, the 'perfect cosmological principle'. The cosmological principle says that the Universe looks much the same wherever you are in it, and that the laws of physics are the same everywhere. The 'perfect' cosmological principle says that the Universe also looks much the same at any time that you view it. Hoyle hated the term and preferred to refer to the 'wide' cosmological principle when writing about the work of Bondi and Gold. Historian Simon Mitton has neatly summed up the difference between the two approaches: Bondi and Gold started with philosophical principles and tried to find a model to match; Hoyle started with equations and developed a model (or models) on the basis of the equations.

All of this set the scene for a debate in the 1950s between the rival Big Bang and steady-state ideas, which started the decade on an equal footing (if anything, with the steady-state idea more widely accepted). Happily, there was a way to test which of them was a better description of the real Universe. If the steady-state idea was right, the number of galaxies in a certain volume of space (the number density) should be the same at all times. If the Big Bang idea was correct, the number density would be greater in the past. And since light travels at a finite speed, when we look out into the Universe we are in effect looking back in time. The question then was whether the number density of galaxies was greater at greater distances from us, corresponding to galaxies seen longer ago. This is referred to as seeing those galaxies at greater look-back times.

Just at that time, following the development of radar in the Second World War, radio astronomy was being developed. It was discovered that some galaxies emit far more energy at radio wavelengths than as visible light. This meant that they could be 'seen' much farther away than visible galaxies. There was no way to measure the distances to these otherwise invisible galaxies, but as a rule of thumb it seemed reasonable to guess that fainter radio galaxies are farther away than bright radio galaxies, in the same way that Hubble had reasoned that fainter optical galaxies are farther away than brighter optical galaxies.

This counting of faint radio galaxies was pioneered by a group of radio astronomers at Cambridge, headed by Martin Ryle. Hoyle and Ryle were not friends (far from it), and Ryle made certain that the theorists knew nothing of this work until he was ready to publish.

He went public in a lecture in Oxford in 1955, where he said 'there seems no way in which the observations can be explained in terms of a steady-state theory'. But the conclusion was premature. In the same year, Australian radio astronomers reported that their number counts did indeed match the predictions of the steady-state model. The Cambridge study was wrong,* and Ryle's eagerness to rubbish Hoyle had led him into overconfidence. A much bigger survey, using improved telescopes with better resolution to reach out farther into the Universe and farther back in time would be needed to settle the matter. As the debate rumbled on, results from such surveys started to come in during the early 1960s, gradually (but not decisively) tilting the balance against the steady-state idea. But all of this paled into insignificance (at least as far as the Big Bang versus steady-state debate was concerned) with the discovery of the cosmic microwave background radiation. This is (almost) where we came in at Part Zero; it's time to bring the story up to date.

---

* Their radio telescope had been unable to resolve close pairs of sources in the sky, so some objects that they counted as single galaxies were in fact two.

8 13.8

Surveys and satellites

Although the discovery and identification of the cosmic micro-wave background radiation fired enthusiasm for the Big Bang model, in the mid-1960s there was still considerable doubt about the value of the Hubble constant, and an apparent conflict between the estimated age of the Universe and the calculated ages of the oldest stars. Gradual improvements in observations over the following years slowly eased the situation, but it would be 30 years before new technology led to a breakthrough as profound as the one made by Hubble and Humason. That new technology was the Hubble Space Telescope (HST), which gave astronomers an unprecedented view of the Universe and completed the calibration of the Cepheid distance scale started by Hubble himself. The importance of this aspect of the mission is highlighted by the proclaimed 'Key Project' of the Hubble Telescope team – to determine the value of the Hubble constant to an accuracy of 10 per cent or less, using essentially the same traditional techniques that Hubble had used. Everything else, including the beautiful pictures that made the general public aware of the HST, was of secondary importance.

The culmination of a tradition

The declared aim of the Key Project team was to use measurements of Cepheid distances to about twenty galaxies (with dozens of Cepheids being studied in each target galaxy) to calibrate the distance scale and work out the value of H. This was a slow process.

Each Cepheid had to be studied at two different wavelengths, to make it possible to allow for the dimming effects of cosmic dust, and each observation took two orbits of the Earth (more than three hours). Then, the observations had to be repeated over weeks or months to get the period of a single Cepheid. Slowing things down even further, the whole project was delayed when it was discovered after the launch of the HST in April 1990 that there was a fault in the optics of the telescope. This could not be repaired until a manned mission to the satellite in December 1993, so the Key Project work only really began in 1994. As the early results started to come in and were shared with the astronomical community (without yet providing an accurate determination of H), they stimulated discussion among astronomers about the distance scale of the Universe, which led (among other things) to my own modest contribution to the story.

Those early results from the Key Project suggested a rather high value for H, about 80 in the usual units, but with estimated errors of about 20 per cent, making any value from 64 to 96 possible. Recalling Eddington's comments about the mediocrity of the Milky Way (see page 171), I confidently assured my colleagues at the University of Sussex, following a seminar on the subject where the speaker favoured the high end of this range, that the correct value must actually lie at the bottom of this range if the Milky Way is just an average-sized spiral. The bigger the value of H, the closer other galaxies must be, so the smaller the size required to explain their appearance in the sky, making our home a continent among islands.

To my surprise, two colleagues, Simon Goodwin and Martin Hendry, took up the challenge implied by my throwaway remark. They said that we could, first, use HST data and other observations to check whether the Milky Way really is an average spiral, and then, if it did turn out to be average, use that fact to work out a value for H. I have explained the details in *The Birth of Time*, but the process is simple to understand in outline. Looking first at seventeen nearby spiral galaxies with accurately determined distances (some from ground-based observations, some from the HST) and working out their actual sizes from the sizes they appear in the sky, we found that our Galaxy is very slightly smaller than the average (a Milky Way diameter of 26.8 kiloparsecs compared with an average of 28.3 kiloparsecs). Eddington, we hoped, would have approved. Then, we took data from a catalogue known as RC3 (from *Third Reference Catalogue of Bright Galaxies*), which gave us redshifts for 3,827 spirals. With a bit of computer modelling, we did the equivalent of adjusting the value of H to slide all these galaxies in and out relative to the Milky Way, until we found a value where the average size of the thousands of galaxies was the same as the average size of our local sample of seventeen (plus the Milky Way itself). The value for H that we came up with, published in 1997, was about 60. In truth, this is a pretty rough and ready technique, and the most significant aspect of the work was that it showed that, statistically speaking, there is only a one in twenty chance that H could be bigger than 75. What we had really found was that the lower end of the early Key Project result was indeed more likely to be correct.

As the Key Project rolled on, this is exactly what emerged from their more accurate measurements, made using observations of more galaxies. The final results from the HST Key Project were published in 2001, based on Cepheid measurements and other objects calibrated using Cepheid distances out to 400 million parsecs (Mpc). The other distance indicators calibrated using Cepheid distances included supernovae. Their bottom line was that the Cepheid distances alone implied a value for H of 71, plus or minus eight, while including other measurements, including the supernova data, gave a value of 72 ± 8. This was the culmination of the project initiated by Hubble more than 70 years earlier; it was, essentially, the end of the road for the traditional technique for determining H (and therefore the age of the Universe) by measuring distances to galaxies and comparing them with redshifts. There would be small refinements in the early 21st century, but nothing spectacular.*

You may, though, have noticed something disturbing about the number. In an Einstein–de Sitter universe, a value for H of 72 implies an age of about 9 billion years – much less than the ages of the oldest stars as understood at the time. By 2001, it was clear that we do not live in an Einstein–de Sitter universe. The evidence had come from a variety of sources, not least satellite observations of the cosmic microwave background radiation.

Too perfect?

The first satellite sent aloft to study the background radiation was the Soviet RELIKT-1, launched in 1983. This established the viability of such missions, but was not sensitive enough to reveal any variations in the radiation from place to place in the sky. This was important, because if the radiation really did come from the Big Bang, it should bear the imprint of the fluctuations in the early Universe which had grown to become the galaxies we see today. By the early 1980s, cosmologists were mildly concerned that the background radiation seemed to be too smooth to allow for such irregularities and that the flatness of the Universe – the way it is balanced between expanding forever and re-collapsing – seemed to be too good to be true. The critical density required to make the Universe flat changes as time passes (it is different in different cosmic epochs). Einstein's equations tell us that if a universe is born in a Big Bang with a tiny bit greater density than the critical density required to make it exactly flat, then the difference from flatness will grow as time passes, because the pull of that extra matter slows the

---

* By 2009, these refinements had pinned the value of H down to 74.2 plus or minus 3.6, that is, in the range from 70.6 to 77.8.

expansion and keeps the density high. Conversely, if a universe starts out with a tiny bit less than the critical density, the difference will grow in the other direction because the universe can now expand more easily, spreading matter thinner and thinner as time passes. Absolute flatness is the least likely of all the possibilities.*

Although people were aware of the problem before, nobody really bothered much about it until Robert Dicke and Jim Peebles, two of the Princeton researchers involved in the identification of the background radiation in the mid-1960s, drew attention to it at the end of the 1970s. In order to explain the flatness of the Universe today, it turns out that the density at the time of the Big Bang must have been within one part in 10^{15} (one in a million billion) of the critical density at that time. It was clear that this must be telling us something fundamental about the birth of the Universe, but in 1979, nobody knew what – at least, not until 6 December that year. Alan Guth, a young researcher at Cornell University, had heard Dicke talk about the flatness problem in a lecture at Cornell in the spring of 1979. Intrigued by the puzzle, he tucked it away at the back of his mind and read up on cosmology whenever he had the chance. In October, he moved to Stanford to work at the Linear Accelerator Center for a year. His knowledge of particle physics began to gel with what he had learned about cosmology, and on 6 December, after discussing the puzzle with a visitor from Harvard, Sidney Coleman, something clicked. Working through the night and into the small hours of Friday 7 December, Guth wrote out his big idea in

---

* Late in his career, Alan Sandage drew attention to a related puzzle, the horizon problem, describing it at the end of the 1980s as 'the most important problem in the field [of cosmology]'. The horizon problem is the puzzle that the Universe looks the same on opposite sides of the sky (opposite horizons) even though there has not been time since the Big Bang for light (or anything else) to travel across the Universe and back. So how do the opposite horizons 'know' how to keep in step with each other?

a notebook, under the heading 'SPECTACULAR REALIZATION'. He knew he was on to something important.

Guth realised that a process called symmetry breaking – involving a phase transition not unlike the way steam gives out latent heat when it condenses to form water – could have poured out energy in the first split second of time, pushing the Universe through a phase of rapid expansion, which he called inflation, and ending up with the Big Bang. (People often make the mistake of using the term Big Bang to include inflation, but the crucial point is that inflation came *before* the Big Bang). During inflation, the size of the Universe increases exponentially, doubling in size once every hundredth of a trillionth of a trillionth of a trillionth of a second (10^{-38} sec). In this picture, everything we see in the observable Universe today was inflated from a seed less than a billionth the size of a proton to about the size of a basketball within about 10^{-30} sec. (Another analogy is to imagine inflating a tennis ball up to the size of the visible Universe in a comparably short time.) Only then did the Big Bang take over.* The Universe we see around us is so uniform because it has been inflated from a seed so small that there was no room for any variations in density within it. This also resolves the flatness problem, because inflation flattens the Universe in the same way that the surface of an expanding balloon, or any other expanding sphere, is flattened when it is inflated. The surface of a tennis ball, which is essentially a two-dimensional entity wrapped around a third dimension, is very obviously curved; but if that tennis ball were inflated to the size of the visible Universe, and you could somehow move about on its surface, any measurements you could make would show that it was indistinguishably close to being flat,

---

* The idea was developed further by the Russian-born American Andrei Linde, and others, but that story goes beyond the scope of this book. See *In Search of the Big Bang.*

just as the real Universe is (but in three dimensions, not two).* The origin of the 'seed' itself would have been a so-called quantum fluctuation, a tiny disturbance in the fabric of spacetime blown up by inflation before it could disappear.

The icing on the cake is that during inflation more of the tiny disturbances known as quantum fluctuations would have taken place in the embryonic Universe and been stretched by the inflation, leaving ripples in the stuff that became the Big Bang proper. These ripples – often referred to as anisotropies – are the seeds on which structures like galaxies (actually, clusters and superclusters of galaxies) formed, and they should have left an imprint on the background radiation. Working backwards from the size of fluctuations seen in the Universe today, it is straightforward to work out the size of these fluctuations in the background radiation, in terms of how much the temperature deviates from the average in different parts of the sky – about one part in a hundred thousand. So, for a temperature of 2.7 K, the fluctuations are about plus or minus 0.00003 K (or 30 millionths of a degree). Working forward from inflation theory, it is possible to predict the kind of pattern that should be seen, the mark of stretched quantum fluctuations. Inflation should have left a clear signature in the sky, if only we had detectors sensitive enough to see it. It's hardly surprising that RELIKT-1 (there never was a RELIKT-2, incidentally) failed to detect these tiny ripples. But the next satellite to probe the background radiation had more sensitive detectors.

NASA's COBE mission (from Cosmic Background Explorer) was launched in November 1989. The advantage of putting even small

---

* It also solves the horizon problem, because regions of the Universe that are now widely separated used to be in contact but were ripped apart by the superfast stretching of space. This stretching occurs, in a sense, faster than light, but nothing moves *through* space faster than light. Sandage found this compelling evidence for inflation, especially (of course) when it was borne out by observations.

radio frequency detectors in orbit, rather than making observations from the ground, is that it makes it easier to reduce interference from the gas and dust of the Milky Way. This interference is less at shorter wavelengths (right into the infrared), but at these shorter wavelengths water vapour in the atmosphere of the Earth blocks the radiation from reaching the ground. So COBE and other satellites more than make up in sensitivity what they lose in size compared with ground-based telescopes. (For the same reason, observations are also made from mountain tops, from the cold, dry air of the Antarctic, and using instruments flown on high-altitude balloons.)

The first observations from COBE showed that the spectrum of the background radiation is a perfect black-body curve, corresponding to a temperature of 2.725 K. The results were presented at a meeting of the American Astronomical Society on 13 January 1990. When John Mather, the originator of the COBE project, unveiled a slide showing the spectacular agreement between theory and observation, the audience burst into spontaneous applause. But this was only a start.* The hard work was just beginning.

It took more than a year for the detectors on board the satellite to scan the entire sky, making 70 million measurements with each of three detectors. It then took months for the team to analyse the data and combine all the measurements into a map of the sky showing how the temperature of the background radiation differs slightly from one place to another. So, it was in 1992 that they announced that there really are tiny differences in the temperature of the background radiation from one place to another: ripples in the background radiation, with the 'hot' spots in the sky just 30 millionths of a degree hotter than average, and the 'cold' spots 30 millionths of a degree cooler than average. These differences are the same on

---

* In 2006, Mather and his colleague George Smoot shared the Nobel Prize for the COBE work.

all scales – large hot spots are no hotter than small hot spots, and so on. This exactly matched the predictions of the kind, and size, of fluctuations that would have been imprinted on the Universe during inflation, providing evidence of the presence of slight differences in the density of matter in the early Universe: the irregularities from which clusters of galaxies grew. The Universe was not too 'perfect' after all. What else could studies of the background radiation reveal? The success of COBE inspired a series of ground, balloon and space experiments to be devised and constructed to probe more details of the background radiation. But developing such projects takes a long time – Mather's original proposal for what became COBE was made in 1974, just ten years after the discovery of the background radiation and fifteen years before the satellite was launched – and while they were being developed our ideas about the nature of the Universe were also being developed.

The dark side

Astronomers (at least, *some* astronomers) have known since the 1930s that there is more to the Universe than meets the eye. But it was only at the end of the 20th century that they realised just what an insignificant component of the Universe everyday matter (baryonic matter, the stuff we are made of) is.

Back in the 1930s, the Dutch astronomer Jan Oort studied the motion of stars within the Milky Way and found evidence for the presence of a lot more matter than we can see in the form of bright stars. Stars like the Sun, moving in roughly circular orbits around the centre of the Galaxy, in the disc of the Milky Way, bob up and down slightly in their orbits, moving out of the plane a little way before being tugged back in towards it. The motion of an individual star cannot be studied over thousands of years, but, as ever, the statistics of the distribution of stars and their velocities can be used to reveal what is going on. What is going on is that patterns of movement of the bright stars show that they are being affected by the gravitational influence of unseen dark matter, as well as by the gravity of all the other bright stars. Nobody was too worried about this in the 1930s, since it was natural to expect there to be plenty of dust and gas between the stars. Indeed, we now know that this kind of dark matter, made of the same sort of stuff that we are made of ('baryonic matter' essentially means atomic matter – anything made of protons, neutrons and electrons) contributes about as much mass overall as all the bright stars in the Galaxy. But that still doesn't entirely account for the way the stars of the Milky Way move.

On a much larger scale, the Swiss astronomer Fritz Zwicky, working at Caltech in the 1930s, found evidence of the presence of dark matter by studying whole clusters of galaxies. You can estimate the mass of a galaxy from its brightness, which depends on the number of stars it contains. So, you can estimate the mass of a cluster by adding up the masses of all the galaxies it contains. The speed with which individual galaxies in a cluster are moving relative to one another can be measured using the Doppler effect (with the cosmological redshift for the cluster subtracted out). Zwicky pointed out that in many clusters the galaxies are moving faster than the escape velocity for the cluster – too fast for the cluster to be held together by its own gravity. The speeding galaxies should have flown apart and 'dissolved' the cluster long ago, when the Universe was young, unless there was a lot of unseen dark matter in the cluster as well, holding them in its gravitational grip. Again nobody (except Zwicky!) worried too much about this, right up into the 1960s. Even when I was a student, Zwicky was regarded as a bit of a crank with a bee in his bonnet as far as dark matter was concerned, although he had a fine reputation for other work. Until the 1960s the Big Bang model wasn't fully established anyway. In addition, it always seemed possible to invoke the presence of unseen but ordinary objects – faint stars called brown dwarfs, clouds of gas or hordes of Jupiter-like planets. But things began to change when the Big Bang idea became established and especially after Hoyle and his colleagues worked out the details of Big Bang nucleosynthesis, in the second half of the 1960s.

The amount of helium and deuterium manufactured in the Big Bang is related to how hot it was (which we know from measurements of the background radiation), the density of baryons at the time, and how fast the Universe was expanding and cooling while nucleosynthesis was going on. Turning this around, by measuring the proportions of these light elements in stars today (no easy task!), we can work out the density of baryons in the Big Bang. When these calculations were carried through, they indicated that the density of baryonic matter is much less than the critical density needed to make the Universe flat. At the time, this was seen as evidence that the Universe is open and will expand forever. Many cosmologists were horrified at the thought that there might be another kind of matter, non-baryonic matter, which dominated the behaviour of the Universe. But from the 1970s onwards, the evidence built up that this is indeed the case. Studies of the rotation of other spirals showed that they are all held in the grip of halos of dark matter. Computer simulations of how galaxies form in the expanding Universe showed the need for a great deal of dark matter to provide the gravitational 'valleys' into which baryons could flow, like streams running down the sides of a valley on Earth, to form stars and galaxies. Without this material, which became known as cold dark matter,* the baryons would have been dispersed by the expansion of the Universe and never clumped together to form the bright objects we see around us (or, indeed, to form us). There was also a growing weight of evidence, plus the prediction from inflation, that the Universe is

---

* The exact nature of cold dark matter remains a subject of intense debate and research, beyond the scope of this book. But you can think of it as being a sea of particles filling the Universe, which interact with baryonic matter only by gravity.

indeed flat. By the middle of the 1980s it was clear that dark matter dominates the Universe, and that not only has at least 90 per cent of everything never been seen, at least 90 per cent of everything is not even made of the same sort of stuff that we are made of.

But then it turned out that even cold dark matter (CDM) is insufficient to explain the appearance of the Universe. We do not need to go into all the details. One key piece of evidence will suffice to set the scene for the discovery that drew the attention of a wide audience, not just cosmologists, to the implications of all this; it goes by the name of the 'baryon catastrophe'.*

The baryon catastrophe refers to this puzzle: studies of the amount of hot gas in clusters of galaxies suggest that the proportion of baryons to dark matter in the Universe is too great to allow the possibility of there being exactly the right amount of all kinds of matter put together to match the predictions of the simplest versions of inflation and make spacetime flat.

It has become firmly established that most of the matter of the Universe is in some invisible form. Theorists delight in playing with mathematical models that include exotica with names such as Cold Dark Matter, Hot Dark Matter, WIMPs and Mixed Dark Matter; but the observers have slowly been uncovering an unpalatable truth. Although there is definitely some dark matter in the Universe, there may be less in the Universe than some of these favoured models imply. The Universe may not solely consist of matter, regardless of how exotic.

The standard model of the hot Big Bang (incorporating the idea of inflation, which invokes a phase of extremely rapid expansion during the first split-second of the existence of the Universe) says that the Universe should contain close to the 'critical' amount of

---

* This explanation of the baryon catastrophe is adapted from my book *Companion to the Cosmos*, published in 1996.

matter needed to make spacetime flat and to just prevent it from expanding forever. But the theory of how light elements formed in the early Universe (primordial nucleosynthesis) limits the density of ordinary baryonic matter (protons, neutrons and the like) to about one twentieth of this critical amount. The residue, the vast majority of the Universe, consists (in the standard picture) of some kind of non-baryonic matter: exotic particles with names such as axions. These particles have never been seen directly, although their existence is predicted by the standard theories of particle physics. In the favoured cold dark matter (CDM) model of the Universe, the gravitational influence of the dark particles on the bright stuff gives rise to structures as the Universe evolves.

The evidence for dark matter comes from observations on a range of scales. Within our own Galaxy, the Milky Way, there is at least as much unseen matter as there is matter in visible stars. Observations of gravitational lensing of stars in the Magellanic Clouds suggest that this particular component of the dark matter may be baryonic: either large planets or faint, low mass stars known as brown dwarfs. There is also evidence from the speed at which stars and gas clouds orbit the outer parts of disk galaxies for more extensive halos of dark matter, but once again these could be baryonic. As far as individual galaxies are concerned, there is actually no need to invoke CDM at all.

There is no reason to suppose, however, that the contents of galaxies are representative of the Universe as a whole. When a protogalaxy first collapsed, it would have contained the universal mix of baryonic matter (in the form of a hot, ionised gas) plus dark matter. The dark matter is 'cold' in the sense that individual particles move slowly compared with the speed of light, but, like the baryonic stuff, they have enough energy to produce a pressure which keeps them spread out over a large volume of space. The baryons lose energy by radiating it away electromagnetically, so they cool very quickly;

the baryon component of the cloud loses its thermal support and will sink into the centre of the protogalactic halo to form the galaxy that we see today. This leaves the dark matter, which cannot cool (because it does not radiate electromagnetically), spread out over a much larger volume.

So to find a more typical mixture of material we must look at larger, more recently formed structures, in which cooling is less efficient. These are clusters of galaxies. A typical rich cluster may contain a thousand galaxies. These are supported against the attractive force of gravity by their random speeds, which can be more than a thousand kilometres per second, and are measured from the Doppler effect produced by their motion, which shifts features in their spectra either towards the blue or towards the red. (This is independent of the redshift produced by the expansion of the Universe, which has to be subtracted out from these measurements.) By balancing the kinetic energy of the galaxies against their gravitational potential energy, it is possible to estimate the total mass of the cluster. This was first done by Fritz Zwicky in the 1930s and led to the then surprising conclusion that the galaxies comprise only a small fraction of the total mass. This was so surprising that for several decades many astronomers simply ignored Zwicky's findings.

Without the experimental background in particle physics, or the cosmological models which are available today, it would have been natural for those astronomers who did take the observations seriously to identify this missing matter as hot gas. This was not done, perhaps because the physical condition of the gas would render it undetectable by any means available at the time. The gas particles are moving at similar speeds to the galaxies, which is equivalent to a gas temperature of about 100 million degrees; this is sufficient to strip all but the most tightly-bound electrons from atomic nuclei, leaving behind positively-charged ions. Such an ionised gas emits mainly at X-ray energies, which are absorbed by the Earth's atmosphere. It was only with the launch of X-ray satellite observatories in the 1970s that clusters were found to be very bright X-ray sources and it was finally realised that the hot gas, or intracluster medium (ICM), cannot be neglected.

The ICM has turned out to be a very important component of clusters of galaxies. Not only does it contain more matter than is present in the galaxies, but its temperature and spatial distribution can be used to trace the gravitational field and, hence, the total mass of the cluster in a much more accurate way than from the galaxies alone. To obtain the total mass of gas, one looks at the radiation rate. This radiation is produced in collisions between oppositely-charged particles (ions and electrons) and so depends upon the square of the gas density. We observe only the projected emissions, as if the cluster were squashed on the plane of the sky, but, assuming spherical symmetry, it is relatively easy to 'invert' this to find the variation of density with distance from the centre of the cluster. The gas is found to be much more extended than the galaxies and can, in some cases, be traced out to several million light years from the cluster centre. Although the galaxies dominate in the core of the cluster, there is at least three times as much, and probably a lot more, gas in the cluster as a whole as there is matter in the form of galaxies. (It is not the mass of gas which is uncertain but the mass of the galaxies.) But even the combined mass of gas and galaxies is less than the total cluster mass, showing that a large amount of dark matter is also present. The hot gas is supported against gravitational collapse in the cluster by its pressure. To work out details of how the pressure falls off with distance from the centre, we would have to know the variation of temperature with distance from the cluster centre. It is usually supposed that the gas is isothermal (the same temperature right across the cluster). This is consistent with both observations and numerical simulations, which show little variation of either random galaxy speeds or gas temperature across the cluster. It is possible that the gas temperature may fall in the outer parts of clusters; this would tend to lower the overall mass estimates.

A study by David White and Andy Fabian of the Institute of Astronomy in Cambridge, published in 1995, examined data from the Einstein satellite for nineteen bright clusters of galaxies. They compared the mass of gas with the total cluster mass and concluded that the gas comprises between 10 and 22 per cent of the total cluster mass, with an average value of about 15 per cent. These fractions would increase by between 1 and 5 per cent (of the total mass) if the mass of galaxies were included. So, the total baryon content of clusters is much greater than the 5 per cent predicted by the standard CDM model for a flat Universe. You still need some dark matter (to the relief of the particle physicists), but only five times as much as there is baryonic matter, not twenty times as much. Since the Big Bang models still say that only 5 per cent of the critical density can be in the form of baryons, and if the distribution of matter in clusters of galaxies is typical of the Universe at large, this means that overall there can only be about 30 per cent of the critical density, even including the dark stuff. In other words, five times more CDM than baryons. If you want to keep the high overall value of the density parameter, you have to allow much more than 5 per cent of the total mass of the Universe to be in the form of baryons, but this is forbidden by the rules of primordial nucleosynthesis.

What is the resolution of this problem? At the time I wrote *Companion to the Cosmos*, there was still a lively debate among astronomers about the exact value of the Hubble constant. In the calculation I have just presented, I assumed a Hubble constant of 50 kilometres per second per megaparsec, at the lower end of the accepted range, corresponding to a large, old universe.

In the cosmological models, as the Hubble constant is lowered, the calculated baryon fraction increases. But the predicted baryon fraction from primordial nucleosynthesis increases even faster, and so the discrepancy between the two is reduced. By making the Hubble constant low enough, one could reconcile the two, but long before this happens the baryon fraction becomes one. Since more than 100 per cent of the mass of the Universe cannot be in the form of baryons, this argument can be reversed to place an absolute lower bound on the Hubble constant of about fourteen, in the usual units. Even in 1996 nobody I knew would countenance going to such extremes.

So, one of the cherished foundations of the then-standard model had to be relinquished. Perhaps the least fundamental of these is that the dark matter must be 'cold'. Hot dark matter, made of particles (such as neutrinos) that emerge from the Big Bang with speeds close to that of light, is unable to cluster efficiently due to the large random motions of its particles. At first sight, you might guess that it could fill the space between clusters of galaxies with huge amounts of matter, so that even the clusters are not representative of the stuff of the Universe. However, hot dark matter cannot comprise more than about one-third of the total amount of dark matter, because interactions between the hot stuff and ordinary baryonic matter would slow the development of structures such as galaxies and clusters, delaying their formation until later times; this conflicts with the observed number of distant, old radio galaxies and quasars.

People have also toyed with the idea of nonstandard nucleosynthesis, for example, allowing the baryon abundance to vary from place to place. This allows some relaxation of the upper bound on the baryon fraction, but the models are rather contrived, and anyway the models do not work as well as the standard one.

We were left, in the mid-1990s, with two possible, simple explanations, one of which was that the mass density of the Universe is much less than the critical density. If 'what you see is what you get', the Universe could contain as much as 5 per cent baryonic material, in terms of the mass density, and about 30 per cent of the critical density in the form of all kinds of matter put together (so, roughly five times as much dark matter as baryonic matter). The baryons themselves would be mostly in the form of about a third hot cluster gas and about two-thirds in the form of galaxies. The rest of the stuff of the Universe would be mainly cold dark matter, perhaps with a smattering of hot dark matter. The Hubble constant could then be rather higher than 50, in line with the Hubble Key Project and COBE measurements. But it would mean that the Universe is not flat, in conflict with the idea of inflation.

That left one possibility, which theorists were aware of but were reluctant to promote. As I wrote in 1996: 'If cosmologists then wish to preserve the idea of a spatially flat Universe, as predicted by theories of cosmic inflation, then they may have to reintroduce the idea of a cosmological constant.' The ink was scarcely dry on that pronouncement when evidence for the existence of the cosmological constant emerged in dramatic fashion, surprising its discoverers, who knew nothing of these cosmological predictions.

Supernovae and superexpansion

The two teams that made the discovery were surveying the distant Universe by studying supernovae seen at very high redshifts. Earlier studies of galaxies close enough to us for their distances to be determined by other methods had calibrated the brightness of a particular class of supernova, known as Sn1a, that have approximately the same intrinsic brightness. This is thought to be because they form from white dwarf stars that gradually accrete matter (probably from a nearby companion in a binary star system), getting heavier and heavier until they reach a very precise critical mass (the same critical mass in every case, whatever mass they started out with). At this point the pressure inside the star triggers a runaway nuclear reaction which releases energy that blows the outer layers apart and briefly makes the single star shine as brightly, in round numbers, as 4 billion Suns. For cosmology, just how the supernova is triggered is less important than the fact that Type 1a supernovae each have about the same brightness,* so they can be used as standard candles. The apparent brightness of a Type 1a supernova tells us how far away it is, and this, of course, can be compared with its redshift.

In the second half of the 1990s two large teams of researchers (each including dozens of members working in different places around the world) were pushing the available technology (better telescopes on the ground and in orbit, better detectors in the form of CCDs, and more powerful computers) to map the distribution of the faintest and most distant supernovae they could find, and therefore to map the distribution of the galaxies in which those supernovae reside. Because astronomers use the letter z for redshift, one team called themselves 'The High-z Supernova Search Team'. Their leaders were Brian Schmidt, of the Australian National University, and Adam Riess, of Johns Hopkins University. Their friendly rivals were a group dubbed 'The Supernova Cosmology Project', headed by Saul Perlmutter, of the University of California, Berkeley. The results of these surveys were so dramatic that it was just as well two independent teams were involved, since the fact that they both found the same thing lent weight to the discovery.

Type 1a supernovae are rare. In a typical galaxy like the Milky Way, only two or three of them occur in a thousand years. But by photographing dozens of patches of sky, each containing hundreds of faint galaxies, the teams could be sure of catching these stellar explosions in the act. If, on average, two supernovae go off in a single galaxy in a thousand years, then if you look at 50,000 galaxies, you would expect to see about a hundred supernova explosions

---

* In fact, a subset of Type 1a supernovae have the same brightness, but these can be distinguished by detailed observations, which we don't need to go into here.

each year. This provides a way to study the Universe at very great distances, which, because of the finite speed of light, means study-ing it as it was long ago. One of the things the researchers expected to use the surveys for was to measure how much the expansion of the Universe has slowed down, as gravity claws back the expansion that was set going in the Big Bang.

Results from the two surveys began to come in in 1998, involving studies of supernovae seen at a time when the Universe was roughly half as old as it is now. To their surprise, both teams found that the galaxies in which distant supernovae reside are not receding from us as rapidly as implied by their redshifts, assuming that the Hubble constant has always had the same value as the one determined from studies of nearer galaxies. This means that the Universe was expand-ing more slowly in the past – the opposite of what they had expected to find.* If it was expanding more slowly, then it must have taken longer to reach its present state. In other words, the supernova data tell us that the age of the Universe must be significantly more than 9 billion years. Just how much more would be revealed by the next generation of satellites.

But if the Universe was expanding more slowly in the past, then, relatively speaking, it is expanding more rapidly today. The expan-sion of the Universe is accelerating. This is the way the discovery was usually presented in the headlines and stories that followed the announcement of the discovery. Something is pushing the Universe outward, with enough strength to (almost certainly) overwhelm gravity and make the Universe expand forever, at an ever faster rate. This 'something' became known as 'dark energy'.

The simplest explanation of the dark energy is that it is a

---

*   It's always a delight when a scientific study produces the opposite of the expected result; it is one way of being sure that no cheating or wishful thinking has been involved!

manifestation of the cosmological constant, lambda (Λ).* If this really is a constant, and has had the same (small) value ever since the Big Bang, there must be the same amount of dark energy in every cubic centimetre of space at all times. 'New' dark energy is created to fill the extra space as the Universe expands. Sound familiar? Mathematically, this is exactly the idea that Einstein abandoned and which Fred Hoyle developed as his C-field cosmology, with Λ as the chosen symbol in the equations instead of C, and with the density of *matter* going down as the Universe expands, even though the dark energy density stays the same.** The dark energy contributes a kind of springiness to space, pushing it outwards, while gravity pulls it back. At first, just after the Big Bang, gravity dominates, because there is not very much dark energy. This slows the expansion of the Universe. But the dark energy density stays the same while the matter density declines, weakening the effect of gravity. The dark energy effect stays the same while the gravity effect gets weaker. At some crucial time, billions of years ago, the dark energy became dominant. From then on, the expansion of the Universe got faster.

There is, though, more to the story. Einstein taught us that matter and energy are equivalent. The presence of dark energy in the Universe acts, gravitationally, like the presence of matter, quite apart from the springiness it gives to space. Roughly speaking, the amount of dark energy needed to explain the observations is, in gravitational terms, about twice as much as the amount of matter (light and dark combined) in the Universe. If the Universe is indeed flat, and about one-third of it is matter, with two-thirds being dark energy associated with the cosmological constant, everything fits and there is no

---

* As ever, there are more complicated 'explanations' on offer as well, but I see no need for them.

** Imagine what would have happened if the supernova results came before the cosmic microwave background radiation was discovered. This could have been interpreted as evidence for the steady-state cosmology!

baryon catastrophe. In 2011, the leaders of the two teams shared the Nobel Prize for their work. The official announcement of the award said that: 'The discovery [of Λ] came as a complete surprise even to the Laureates themselves.' But it didn't come as a complete surprise to those cosmologists who were already aware of the desirability of invoking a cosmological constant.

The addition of this piece of the puzzle completed, in outline, what has become the standard model of cosmology, known as ΛCDM (or Lambda–CDM), because it includes both the cosmological constant and cold dark matter. In the first fifteen years of the 21st century, the story of cosmology became the story of the (successful) attempt to pin down the parameters of the standard model, including the age of the Universe. This was largely achieved by experiments on board two satellites, known as WMAP and Planck.

Sounding out the Universe

It is a mind-blowing concept (even to someone who, like me, has grown up with the idea) that only about 5 per cent of the Universe is in the form of baryonic material, the stuff we are made of, with maybe 25 per cent in the form of cold dark matter, like nothing we have ever seen, and the rest in the form of dark energy. I'm sometimes asked why nobody noticed any of this before, and the best explanation I can come up with is to spell out just how little dark energy there is in every cubic centimetre of space.

Matter is not spread out uniformly across the Universe, but forms clumps, in which things like galaxies, stars and people form. But the Λ field *is* spread out evenly, with an energy equivalent to a mere 10^{-27} grams in every cubic centimetre – not just every cubic centimetre of 'empty' space, but every cubic centimetre of everything. There is no way that this could be detected in any laboratory experiment using present-day technology. The amount of dark energy contained in the volume of the entire Earth is about as much as the amount of electricity used by the average US citizen each year at the beginning of the 21st century. A sphere as large across as the Solar System out to the orbit of Uranus only contains the same amount of dark energy as the amount of electromagnetic energy (heat and light) radiated by the Sun every couple of hours. To see the influence of the cosmological constant at work, you really do have to look at the Universe at large, and that is where the satellites come in.

The observations of the microwave background tell us about the time when the Universe became transparent and electromagnetic radiation was able to stream freely through space. This happened a bit less than 400,000 years after the split-second during which inflation occurred. Before then, the Universe was so hot that neutral atoms could not exist, and there was a sea of electrically-charged particles, electrons and nuclei of (mostly) hydrogen and helium, which interact with electromagnetic radiation. Among other things, these interactions imprinted the ripples discovered by COBE; those ripples, the primary anisotropies, were themselves imprinted by inflation, during the first split-second of the life of the Universe. Once the Universe cooled to about the same temperature as the surface of the Sun today (about 6,000 K), electrically-neutral atoms could form and the radiation was free to go on its way. For the same reason, electromagnetic radiation escapes from the surface of the Sun at this temperature, which is why it forms the visible surface of the Sun. In the case of the Universe, the place where this happens is known as 'the last scattering surface'.* The irregularities in the radiation were not, however, solely a result of the anisotropies imprinted by inflation. It was not completely unaffected between the time of inflation and the time of last scattering. The way matter was distributed across the Universe during the first few hundred thousand

---

* The term 'surface' is a bit misleading here. The decoupling probably took place over an interval of about a hundred thousand years, finishing when the Universe was a bit less than 500,000 years old.

years produced a smaller 'signature' in the form of secondary ripples in the background radiation. The effects are small even compared with the one part in a hundred thousand primary anisotropies, but, following the success of COBE, researchers set out to measure them in order to fine-tune our understanding of the origin and evolution of our Universe.

The exact nature of these anisotropies – ripples – depends on a balance between two conflicting effects at work in the expanding Universe. Concentrations of baryons (actually embedded in concentrations of dark matter, but the dark matter does not interact with electromagnetic radiation) pull things together by gravity, and make anisotropies bigger. But as long as the material is hot enough to interact with electromagnetic radiation, fast-moving photons (particles of electromagnetic radiation) tend to smooth out irregularities in the distribution of baryons. The competition between these two effects creates features known as acoustic oscillations (sometimes, baryon acoustic oscillations, or BAOs). They can be thought of as pressure (sound) waves in the material of the early Universe; because of the interplay between matter and radiation, some wavelengths grow while others are smoothed away. The resulting mixture of wavelengths contains a great deal of information about the Universe, if it can be interpreted.

What is needed is a technique for picking out which wavelengths are present in the background radiation, and how big they are; fortunately, astronomers have exactly the tool they need to do this job. Power spectrum analysis is a technique for unravelling the different regular variations that combine to make a complicated pattern. It is pretty much infallible, as long as the complicated pattern really is made up from a mixture of simpler variations. If a chord is played on a guitar, each of the six strings may contribute a different note, producing a superficially complicated pattern of pressure waves, which we hear as a distinctive sound. This sound pattern can be picked up by a microphone, converted into electrical signals, and displayed as a messy looking set of squiggles on the screen of, say, a computer. Power spectrum analysis can take that messy pattern and deconstruct it into the individual notes played by the guitar strings. It can also tell you how loud each of those notes is – how much power there is in each component of the spectrum. The patterns in the microwave background radiation revealed by sufficiently sensitive detectors can be analysed in exactly the same way to tell you which 'notes' were playing when the pattern of acoustic oscillations got frozen out, as matter and electromagnetic radiation decoupled at the time of last scattering.

These oscillations contain a great deal of information. Changing the analogy slightly, by analysing the notes produced by the pipes of a church organ, a physicist could tell a great deal about the structure of the organ (for example, the lengths of the pipes) without ever seeing the instrument. The power spectrum of the background radiation is usually presented as a graph displaying the amount of power on different scales (for different-sized oscillations), with large scales to the left and smaller scales to the right. Peaks in this graph correspond to places (scales) where the oscillations are strong, troughs to places where they are weak. Such a graph shows one large peak, trailing off to the right (smaller scales) in a series of lesser wiggles. The first peak could not be measured even by COBE, although it was pinned down with good precision by instruments carried on balloons and by some ground-based observations in the years following COBE's breakthrough. The position of this peak, determined with good precision by the year 2000, tells us about the curvature of the Universe, and provides the key evidence that the Universe is flat, with all that that implies for the density of the Universe and the existence of dark matter and dark energy. Theorists knew that the ratio of the heights of the first and second peaks tells us how much of the matter is baryonic (independently of arguments such as those based on the 'baryon catastrophe') and that the third peak reveals information about the density of dark matter. But COBE was nowhere near sensitive enough to provide details of the pattern of these peaks, and even the balloon-borne instruments could only provide rough measurements. (Balloon experiments cannot observe the whole sky and cannot stay aloft as long as satellites!) The best prospect was to combine the most accurate measurements of small-scale anisotropies made from the ground and from balloons, but covering only parts of the sky, with surveys of the whole sky measuring the larger-scale anisotropies with great precision. That is where the next generation of satellites came in.

Ultimate truth

The first of these satellites was NASA's Wilkinson Microwave Anisotropy Probe (WMAP, pronounced *double-u map*), launched on 30 June 2001. Originally called the Microwave Anisotropy Probe (MAP), in 2003 the satellite was renamed in honour of David Wilkinson, who had died the previous year. The mission had been proposed in 1995 and got the green light in 1997; the speed with which the proposal was turned into hardware and put into space shows just how important it was, and the way the COBE results had shaken up this kind of research. The WMAP detectors were 45 times more sensitive than those on COBE, and they could 'see' details on an angular scale 35 times smaller than the smallest patches that could be resolved by their predecessors – WMAP could identify features only one-fifth of a degree across, about a third of the size of the full Moon as seen from Earth.* It could also observe in five different wavelengths. It was initially intended that the mission would have a two-year lifetime, but it proved so successful that consecutive

---

* COBE had an angular resolution of seven degrees across the sky: fourteen times larger than the Moon's apparent size. This made COBE sensitive only to broad fluctuations of large size.

extensions were granted, until it had completed nine years of observations. It was then (in 2010) moved to a 'graveyard' orbit where it would not get in the way of future satellites, and there it still remains, going round the Sun fourteen times in every fifteen years.

Even the first year of observations by WMAP was spectacularly successful. Data from the satellite pinned down the age of the Universe as 13.4 ± 0.3 billion years, with a Hubble constant of 72 ± 5, with less than 5 per cent of the mass of the Universe in the form of baryons, and with a total matter content of about 28 per cent of the mass needed to make the Universe flat, implying that 72 per cent of the Universe is dark energy. Overall, the pattern of fluctuations matched the predictions of inflation.*

When the WMAP data were combined with measurements made by other instruments, including the balloon measurements, the cosmological parameters could be pinned down even more accurately, and as the years passed and the WMAP observations continued, the results became more and more accurate. One of the most important of these other measures of the Universe came from surveys of galaxies, which showed the imprint of baryonic acoustic oscillations on the way millions of galaxies are distributed across the sky. After nine years, the WMAP data on their own implied an age of 13.74 ± 0.11 billion years, a value for H of 70.0 ± 2.2, a baryon density of 4.6 per cent, a cold dark matter density of 24 per cent, and a dark energy density of 71 per cent. This determination of H is completely independent of the traditional method based on Cepheids; so, the fact that the two results agree with one another (within the range of possible errors) is dramatic confirmation that the whole ΛCDM package is a good description of the Universe.

---

* I should point out that the analysis which follows is based on the simplest interpretation of the data. It is always possible to develop more complicated scenarios, such as having a cosmological constant which changes with time, but I see no need to go down that path unless we are forced to.

If more evidence is needed, the curvature of space was pinned down to within 0.4 per cent of complete flatness. Adding in data from other kinds of observations changed these numbers very slightly, with the estimate for the age of the Universe being pushed up a bit to 13.772 ± 0.059 billion years. But as WMAP was retired, another satellite, the European Space Agency's Planck mission, was taking over and refining the numbers still further.

Planck – named, of course, after the man who first explained the nature of black-body radiation – was one of a pair of highly successful satellites put into orbit by a single launcher, an Ariane 5 rocket, on 14 May 2009. The first proposals for the Planck satellite were submitted to the European Space Agency (ESA) in 1993 (two years before the WMAP proposal was submitted to NASA), and it took sixteen years to go from those initial proposals to a tested satellite ready for launch. This more cautious approach resulted in a superior satellite using more advanced technology. It had a sensitivity slightly more than three times better than WMAP, meaning it was able to measure temperature differences in the sky as small as a millionth of a degree; it covered a wider range of wavelengths than WMAP, and could measure hot and cold spots as small as one-twelfth of a degree across, twice the resolution of WMAP. So, its results have become the benchmark figures for cosmology until even more sensitive detectors are put into orbit. The other satellite launched on the same rocket, Herschel, made an infrared survey of the Universe. Some of my colleagues at the University of Sussex were members of the Herschel team, and I watched the launch with them over a live video link to a big screen at the university. (I felt it tactful not to mention that I was more interested in Planck than in 'their' satellite.) After various manoeuvres, Planck reached its operating orbit on 3 July 2009, and carried out observations until October 2013 when, with its supply of cooling liquid helium exhausted and low on fuel, it was moved into a graveyard orbit and switched off.

The first detailed results from Planck's all-sky survey were released in March 2013; they improved slightly on the WMAP results. This was when the announcement appeared that the 'age of the universe' is 13.82 billion years, prompting me to start thinking about writing this book. To be precise, at that time the Planck data alone gave an age of 13.819 billion years. With data from galaxy surveys of BAOs and other sources included, the 'best estimate' of the age became 13.798 ± 0.037 billion years. More detailed analysis of the Planck data, completed at the end of 2014, and released in February 2015, implied that the value of the Hubble constant is 67.8 and the age of the Universe is 13.799 billion years. With data from the Sn1a and BAO added in, the value for the Hubble constant is 67.74, and the implied age of the Universe is still 13.799 billion years, with slightly smaller error bars of ±0.021 billion years. So, to one decimal place, the overall 'best' estimate for the age of the Universe became 13.8 billion years, since any remaining uncertainty is in the second decimal place. This number is unlikely to change, and it is worth pausing to take on board the fact that at the end of 2015 the cosmological debate now revolves around small differences in the second decimal place, rather than there being any fundamental disagreements. Planck agrees very well with the WMAP result, but it is even more accurate. It's worth stressing that the agreement is far more important than the difference. Even if you take the numbers at face value and ignore the estimated error ranges, the two sets of observations give figures that differ only by about 100 million years over a timespan of roughly 14 *billion* years! This is an 'error' of less than 1 per cent.

Not everything, though, agrees quite so well. Combining Planck and data from other sources, in 2013, the calculated value for the amount of dark energy in the Universe was 69.2 ± 0.01 per cent (revised in December 2014 to 68.3 per cent), with the overall matter density being 31.5 ± 1.7 per cent (revised to 31.7 per cent, with just under one-sixth of that, 4.9 per cent overall, in the form of baryonic matter), and the corresponding value for *H* being 67.80 ± 0.77 (the 'raw' Planck figure, as of December 2014, was 67.15). This implies that the Universe is expanding slightly less rapidly than we previously thought. Although this just overlaps with the WMAP value at the lower end of the WMAP range of possible errors, and WMAP agrees with the traditional technique at the other end of the WMAP error range, the Planck error range does not overlap with the error range of the latest extension of the traditional technique (as of 2014, the best Cepheid plus Supernova value was 73.8 ± 2.4). This may mean that the ΛCDM package is not quite a perfect description of the Universe. Although François Boucher, a leading member of the Planck collaboration, stressed to us that 'we have no strong evidence for anything that goes beyond basic ΛCDM', this is the kind of thing that might one day force us to look at the complications referred to in the footnote on page 215. It just might be a signal of some deviation from the simplest model that we do not understand. One speculative but plausible possibility is that we may live in a region of slightly lower than average density in the Universe, sometimes called a 'Hubble bubble'. Compared with the entire Universe, even the supernova technique is only covering a small region of space (one reason why the microwave measurements provide a more trustworthy measure of the age of the Universe), and if that region is slightly underdense, matter outside the bubble may be tugging on the galaxies we can see, mimicking the effect of a slightly larger Hubble constant.* I stress that this is a speculation

---

* Amusingly, back in 1999 a group at Sussex, of which I was a member, suggested that such a Hubble bubble could do away with the need for lambda altogether. We concluded that: 'Previously popular cosmologies such as open or critical matter density Universes with no cosmological constant may be acceptable if we live in a local underdensity.' See: Goodwin et al., 'The local to global H_0 ratio and the SNe 1a results', 10 June 1999, arXiv:astro-ph/9906187.

and may be wrong, and also that any tweak to the ΛCDM model that might be required as an alternative to explain the discrepancy will probably be small, but this is just the kind of thing that excites scientists. When all the observations are exactly in line with what our theories predict, life can be a bit dull; when there is a disagreement, we have the excitement of developing new explanations to resolve the discrepancies.

Nevertheless, the discovery that the ages of the oldest stars and the age of the Universe are almost the same, with the stars (crucially) being slightly younger than the Universe in which they live is, indeed, one of the most profound discoveries ever made. It powerfully suggests that both the general theory of relativity and quantum mechanics are correct in some fundamental way, and might one day be unified. Most astoundingly of all, according to Planck, we know the age of the Universe to an accuracy of better than 1 per cent. This, indeed, is the 'ultimate truth' of science, and the ultimate proof that science is the best way to understand how the world works.

Glossary

(See also my book *Companion to the Cosmos*, which is one big glossary!)

Absolute magnitude
The brightness of a star if viewed from a distance of exactly 10 parsecs.

Alpha particle
A 'particle' composed of two protons and two neutrons bound together so tightly that in many circumstances it behaves like a single entity. Identical to the nucleus of a helium atom.

Anthropic principle
The idea that the existence of life in the Universe (specifically, human life) can set constraints on the way the Universe is now, and how it got to be the way it is now.

Antimatter
A form of matter in which key properties, such as electric charge, are the opposite of those in everyday matter. For example, the positively-charged positron is the antimatter counterpart of the negatively-charged electron.

Atom
The smallest component of everyday matter that takes part in chemical reactions. All elements, such as oxygen or iron, are made of

particular kinds of atoms. Each atom is made up of a tiny central nucleus surrounded by a cloud of electrons.

Baryonic matter
The name given to matter like the everyday matter here on Earth, made of protons, neutrons and electrons. Strictly speaking, electrons are not baryons, but their mass is tiny compared with that of protons and neutrons.

Big Bang theory
The well-established idea, confirmed by observations (for example, of the cosmic background radiation) that the Universe as we know it emerged from a hot, dense state.

Binary system
A pair of stars orbiting around one another.

Black body
A hypothetical object that absorbs all electromagnetic radiation that falls on it. A hot black body is a 'perfect' radiator of electromagnetic energy. The Sun and stars are very nearly black bodies.

Black-body radiation
The radiation emitted by a black body.

Cepheid
A kind of variable star that changes brightness in a regular way that enables astronomers to work out its average brightness and, therefore, how far away it is.

Classical physics
The rules and equations that apply to things much bigger than atoms.

Cold dark matter (CDM)
The dominant material component of the Universe, present in about the ratio 5:1 compared with everyday matter. The presence of CDM is revealed by its gravitational influence, but nobody knows exactly what it is.

Cosmic background radiation
Radiation left over from the Big Bang, detectable today in the form of a weak hiss of radio noise coming from all directions in space. This is almost perfect black-body radiation.

Cosmological constant
A number which indicates the amount of dark energy in the Universe. Denoted by the Greek letter lambda (Λ).

Cosmological redshift
A change in the wavelength of light from distant objects caused by stretching as the Universe expands. This shifts features in the light towards the red end of the spectrum.

Critical density
The density for which the spacetime of the Universe is flat. The critical density is equivalent to the presence of about five hydrogen atoms in every cubic metre of space.

Dark energy
A form of energy which fills all of space, detected only by its influence on the way the Universe expands. About two-thirds of the mass-energy of the Universe is in this form.

Dark matter
Material detected only by its gravitational pull, which affects the

way galaxies move and how the Universe expands. There is five or six times more dark matter than there is baryonic matter.

Deuterium
A heavy form of hydrogen, in which each atom has one proton plus one neutron in its nucleus.

Disc galaxy
A system of hundreds of billions of stars, most of them in a flattened disc, where there may be a spiral structure. Our Milky Way is a disc galaxy. Many (but not all) disc galaxies have a spiral structure.

Doppler effect
A change in the wavelength (or frequency) of light caused by motion. For objects moving towards us, the wavelength is squashed (blueshift); for objects moving away the wavelength is stretched (redshift). NB the cosmological redshift is *not* a Doppler effect.

Electromagnetic radiation
Any form of radiation, including light, radio waves and X-rays, that is made up of electricity and magnetism. Described by a set of equations named after James Clerk Maxwell.

Electron
Negatively-charged particle that forms the outer part of an atom.

Element
A substance composed of atoms that each have the same number of protons in their nuclei, and therefore the same number of electrons surrounding those nuclei (which determines the chemical properties of the element). Some of the nuclei may have different numbers of neutrons, making them different isotopes of the same element.

Elliptical galaxy
A large system of stars with no obvious internal structure, with an overall shape like that of an American football.

Flame test
The flame test is a simple way to determine the identity of an unknown substance. A clean wire loop is dipped in the substance (a compound, such as sodium chloride) then held in the flame of a Bunsen burner. The heat of the flame excites the atoms (strictly speaking, ions), causing them to emit visible light with a characteristic colour (yellow in the case of sodium).

Galaxy
(lowercase 'g')
A large island in space containing many stars – up to hundreds of billions of stars like the Sun.
(uppercase 'G')
Our home galaxy, also known as the Milky Way.

General theory of relativity
The theory, developed by Albert Einstein, that describes the relationship between matter and gravity in terms of curved spacetime.

Globular cluster
A densely packed ball of stars which may contain millions of individual stars.

Gravity
The force that makes lumps of matter attract one another. For example, the Earth pulls on your body, but your body also pulls on the Earth. Albert Einstein explained how gravity works using the general theory of relativity.

Hubble constant
Also known as the Hubble parameter, a number which measures how fast the Universe is expanding.

Inflation
The early phase in the development of the Universe when a tiny quantum fluctuation expanded to about the size of a basketball in a tiny fraction of a second.

Ion
An atom (sometimes a molecule) which has lost one or more of its electrons is positively-charged and called an ion. The spectra of ions are correspondingly different (in a way which can be calculated) from those of the 'parent' atoms. It is also possible for an atom to gain an electron and have overall negative charge.

Kelvin temperature scale
Temperature measured from the absolute zero of temperature, –273.15°C. Each degree on the Kelvin scale is the same size as a degree on the Celsius scale, but is written without the 'degrees' sign. So 0°C is the same as 273.15 K, and so on.

Kelvin–Helmholtz timescale
The length of time for which a star like the Sun could continue to radiate energy simply by contracting slowly under its own weight – about 20–30 million years. In the middle of the 19th century, astronomers and physicists puzzled over how the Sun kept itself hot. They realised that if it were entirely made of coal, burning in an atmosphere of pure oxygen, it would be burnt out in less than about 100,000 years, and they suspected, from geological evidence, that the Earth had been warmed by the Sun for much longer than that. Hermann Helmholtz, in Germany, and William Thomson (later

Lord Kelvin) in Britain, independently came up with the same solution to the problem. They showed that simply by shrinking slowly in upon itself, the Sun could shine as brightly as it does today for several tens of millions of years, as gravitational energy was converted into heat.

Kirchhoff's law

At a given temperature, the rate of emission of electromagnetic energy by an object is equal to the rate at which the object absorbs electromagnetic energy of the same wavelength (frequency). This law was first stated by Gustav Kirchhoff in 1859, and proved by him in 1861. It led him to develop, in 1862, the idea of a black body, and black-body radiation, which in turn led Max Planck to introduce the idea of quanta into physics.

Lambda (Λ) field

Another name for dark energy.

Light year

The distance light travels in a year – 9.46 thousand billion km; a measure of distance, not of time.

Magnitude

The brightness of a star, measured by astronomers on a scale named after the English astronomer Norman Pogson. The dimmer a star is, the larger its number on the Pogson scale. For historical reasons, a difference in magnitudes of 5 means that one object is a hundred times brighter (or fainter) than another.

Main sequence star

A star in the quiet prime of its life, like the Sun.

Multiverse model

The idea that our entire observable Universe may be just one 'bubble' in a more extended reality – the Multiverse.

Nebula

In its modern usage, a cloud of gas and dust between the stars. Before it was realised that the objects now known as galaxies lie outside the Milky Way, some of them, like the Andromeda Galaxy, were also labelled nebulae; but as applied to galaxies the term is now obsolete.

Neutron

A neutral particle that is part of the nucleus of an atom.

Neutron star

A collapsed object formed from the remnant of an old star, composed almost entirely of neutrons. A typical neutron star contains a bit more mass than the Sun but is only about 10 km in diameter.

Nova

The sudden brightening of a star which makes it look like a 'new' object in the sky.

Nuclear fusion

The process of fusing light nuclei (in particular, those of hydrogen) into heavier nuclei (in particular, those of helium). This releases energy and keeps stars like the Sun shining.

Nucleon

Generic name for protons and/or neutrons.

Nucleosynthesis

The natural processes which build up heavier elements from lighter ones. A little of this happened in the Big Bang (Big Bang nucleosynthesis) but most of the elements except hydrogen and helium have been manufactured inside stars (stellar nucleosynthesis).

Parallax

The apparent movement of an object across the sky when it is seen from two different points. This can be used to calculate the distance to the object by triangulation. It is easy to see parallax at work. Hold a finger up at arm's length, and close one eye. Now close the open eye and open the one that was closed. Your finger seems to jump sideways compared with the more distant background. In principle, you could work out how long your arm is by measuring the angle across which your finger seems to jump, although there wouldn't be much point in doing this.

Parsec

A measure of distance used by astronomers, equal to 3.2616 light years. A parsec is the distance from which the Earth and the Sun will seem to be separated by an angle of 0.5 arc second.

Photon

A particle of light, or of any electromagnetic radiation.

Planet

A large ball of rock or gas, big enough for gravity to make it round, in orbit around a star.

Principle of terrestrial mediocrity

The idea that we do not occupy a special place in the Universe and that our surroundings are typical of those of a star in a disc galaxy.

Proton
A positively-charged particle that is part of the nucleus of an atom.

Quantum physics
The laws and equations that describe the way small things like electrons and atoms behave.

Red giant
A star in the later stages of its evolution, when it swells up to have a diameter about as large as the diameter of the Earth's orbit today.

Reflecting telescope
A telescope that gathers light and magnifies images using a curved mirror.

Refracting telescope
A telescope that gathers light and magnifies images using lenses.

Singularity
A point with zero volume, or a line with zero width.

Spectroscopy
The technique of analysing the light from an object to reveal its composition. Each element, such as hydrogen or carbon, produces distinctive lines in the spectrum, equivalent to a fingerprint or a barcode. The lines in the solar spectrum were first studied by Josef von Fraunhofer.

Spectrum
The rainbow pattern of coloured light seen when white light is split up using a prism. The range of colours we can see extends from red

through orange, yellow, green, blue and indigo to violet. Red has the longest wavelength, violet the shortest.

Speed of light
The ultimate speed limit for anything moving through space, 299,792,458 metres per second (very nearly 3×10^8 m/s).

Star
A hot ball of gas, many times bigger than a planet, which shines because energy is released by nuclear reactions going on in its interior. The Sun is a star.

Steady-state model
The idea that on the largest scales the Universe always presents the same overall appearance, at any time. The success of the Big Bang model rules out the steady-state idea in its original form, but it may be appropriate in the context of the Multiverse and inflation.

Stellar spectroscopy
The study of the spectra of starlight. In a hot gas, collisions between fast-moving atoms raise electrons to excited states. They then drop down producing emission lines. In a cool gas, the electrons absorb background light and are raised to excited states. Stellar spectra reveal which atoms are involved, and therefore what stars are made of.

Supernova
The explosion and extreme brightening of certain kinds of star, with more than a certain amount of mass, at the end of their lives, when the single star can shine for a short time as brightly as a whole galaxy of stars like the Sun. A supernova leaves behind a remnant in the form of either a neutron star or a black hole.

Tunnel effect

An effect arising from the uncertainty principle of quantum physics which allows particles (such as electrons or alpha particles) to escape from or get in to an atomic nucleus even though classical theory says they have insufficient energy to do so. This is related to the dual wave–particle nature of quantum entities.

Wave–particle duality

The idea, confirmed by experiments, that quantum entities may behave either as waves or as particles, depending on the circumstances. This does not mean that the entities *are* waves, or that they *are* particles; we have no way of knowing what they are, and can only build up a picture of what is going on in a particular experiment by making analogies with things in the everyday world, such as waves and particles.

White dwarf

A kind of dead star. The Sun will end its life as a white dwarf, about as big as the Earth is now. One cubic centimetre of white dwarf matter would have a mass of about 1 ton.

Wien's law

A relationship which gives the temperature of a black body in terms of the wavelength at which it radiates the maximum amount of energy in its spectrum. Named after the German physicist Wilhelm Wien (1864–1928), who was awarded the Nobel Prize, in 1911, for his work on the laws governing the radiation of heat.

Sources and Further Reading

Books marked with an asterisk★ are a bit more technical.

Ralph Alpher and Robert Herman, *Genesis of the Big Bang*, Oxford: Oxford University Press, 2001.

Marcia Bartusiak, *The Day We Found The Universe*, New York: Pantheon, 2009.

Jeremy Bernstein, *Three Degrees above Zero*, New York: Scribner's, 1984.

Marcus Chown, *Afterglow of Creation*, London: Arrow, 1993.

Peter Coles, ed., *The New Cosmology*, Cambridge: Icon Books, 1998.

Auguste Comte, *Cours de Philosophie Positive: La Philosophie Astronomique et la Philosophie de la Physique*, vol. 2, Paris: Mallet-Bachelier, Imprimeur-Libraire, 1835.

Ken Croswell, *The Alchemy of the Heavens*, New York: Anchor, 1995.

★Arthur Eddington, *The Internal Constitution of the Stars*, Cambridge: Cambridge University Press, 1926.

John Farrell, *The Day Without Yesterday: Lemaître, Einstein and the Birth of Modern Cosmology*, New York: Basic Books, 2005.

Pedro Ferreira, *The Perfect Theory*, London: Little, Brown, 2014.

George Gamow, *The Birth and Death of the Sun*, New York: Viking, 1940. (Revised and updated as *A Star Called the Sun*, New York: Viking, 1964.)

George Gamow, *The Creation of the Universe*, New York: Viking, 1952.

George Gamow, *My World Line*, New York: Viking, 1970.

Douglas Gough, ed., *The Scientific Legacy of Fred Hoyle*, Cambridge: CUP, 2005.

John Gribbin, *Companion to the Cosmos*, London: Weidenfeld & Nicolson, 1996.

John Gribbin, *In Search of the Big Bang*, revised edition, London: Penguin, 1998.

John Gribbin, *In Search of the Multiverse*, London: Allen Lane, 2009.

John Gribbin, *Einstein's Masterwork: 1915 and the General Theory of Relativity*, London: Icon Books, 2015.

John Gribbin and Mary Gribbin, *How Far is Up?*, Cambridge: Icon Books, 2003.

Fred Hoyle, *Home is Where the Wind Blows*, Mill Valley, CA: University Science Books, 1994.

George Johnson, *Miss Leavitt's Stars*, New York: Norton, 2006.

Alan Lightman and Roberta Brawer, *Origins*, Cambridge, MA: Harvard University Press, 1990.

John Mather and John Boslough, *The Very First Light*, New York: Basic Books, 1996.

Simon Mitton, *Fred Hoyle*, London: Aurum Press, 2005.

Harry Nussbaumer and Lydia Bieri, *Discovering the Expanding Universe*, Cambridge: CUP, 2009.

Dennis Overbye, *Lonely Hearts of the Cosmos*, London: Macmillan, 1991.

Michael Rowan-Robinson, *The Cosmological Distance Ladder*, New York: Freeman, 1985.

George Smoot and Keay Davidson, *Wrinkles in Time*, New York: Little, Brown, 1993.

*Michael Way and Deirdre Hunter, ed., *Origins of the Expanding Universe: 1912–1932*, San Francisco: Astronomical Society of the Pacific, 2013.

End Notes

1. Rhodri Evans's *The Cosmic Microwave Background: How It Changed Our Understanding of the Universe* (Springer, 2015) is the best guide to the history of the investigation of this radiation.
2. See Chown, *Afterglow of Creation*.
3. See Chown.
4. Nobel Lecture
5. November 1948.
6. See Alpher and Herman, *Genesis of the Big Bang*.
7. See Chown.
8. See Mather and Boslough, *The Very First Light*.
9. Reprinted in the collection *Observing the Universe*, edited by Nigel Henbest, Oxford: Blackwell, 1984.
10. See *Nature*, vol. 65 (1902): 587.
11. *Macmillan's Magazine*, 5 March 1862.
12. For sources, see J. Burchfield, *Lord Kelvin and the Age of the Earth*, London: Macmillan, 1975.
13. Quoted in *Rutherford at Manchester*, edited by J.B. Birks, Manchester: Heywood & Co., 1962.
14. Quoted by Burchfield.
15. See *Essential Astrophysics* by Kenneth Lang, Heidelberg: Springer, 2013.
16. See Mitton, *Fred Hoyle*.
17. See Mitton.
18. See Croswell, *The Alchemy of the Heavens*.
19. You can find this in Croswell's book or my own: *Stardust*, London: Allen Lane, 2000.
20. Available in a facsimile edition, edited by Michael Hoskin, London: Macdonald, 1971.
21. See Simon Goodwin, John Gribbin and Martin Hendry, 'The relative size of the Milky Way', in *The Observatory*, vol. 118 (1998): 201–8.
22. See Way and Hunter, ed., *Origins of the Expanding Universe: 1912–1932*.
23. Translation by Ari Belenkiy, in Way and Hunter.
24. See Nussbaumer and Bieri, *Discovering the Expanding Universe*.
25. Quoted by John Farrell, in Way and Hunter.

26. For a discussion of why this should be so, see my book *In Search of the Multiverse*.
27. See Lightman and Brawer, *Origins*.

Index